Flood Response and Crisis Management in Western Europe

Springer
Berlin
Heidelberg
New York
Barcelona
Budapest
Hong Kong
London
Milan
Paris
Santa Clara
Singapur
Tokyo

Uriel Rosenthal, Paul 't Hart (Eds.)

Flood Response and Crisis Management in Western Europe

A Comparative Analysis

With 16 Figures and 8 Tables

Springer

Editors:

Professor Dr. Uriel Rosenthal
Dr. Paul 't Hart
Crisis Research Centre
Leiden University
P.O. Box 9555
2300 RB Leiden
The Netherlands

Library of Congress Cataloging-in-Publication Data
Flood response and crisis management in Western Europe : a comparative analysis /
Uriel Rosenthal, Paul 't Hart, eds. p. cm.
Includes bibliographical references.

ISBN-13: 978-3-642-71999-8 e-ISBN-13: 978-3-642-71997-4
DOI: 10.1007/978-3-642-71997-4

1. Floods--Europe, Western--Case studies. 2. Disaster relief--Europe, Western--Case studies. 3. Emergency management--Europe, Western--Case studies. 4. Crisis management in government--Europe, Western--Case studies. I. Rosenthal, Uriel, 1945 . II. Hart, Paul 't.
HV609.F54 1998 363.34'938'094--dc21 97-43394 CIP

Softcover reprint of the hardcover 1st edition 1998

Typesetting: By Thea H. M. de Beer, Leiden / The Netherlands
Cover: Design & Production, Heidelberg, source of cover figure: THW-Hilberath
SPIN:10556215 32/3020-5 4 3 2 1 0 - Printed on acid-free paper

PREFACE

Throughout the last 20 years, the study of disasters has developed into a well-respected body of knowledge in the social sciences. Disaster research has found its way to several disciplines and it has been remarkably receptive to multi- and interdisciplinary impulses. Many disaster studies show how important and stimulating it is to explore events and developments which reach beyond the usual business of the social and political fabric. There is quite a difference between the daily routines of social and political life, and the upheaval, stress, and shock of a life-threatening danger which may fall upon society. Even when the danger does not materialize to the fullest and society gets off with a fright, the population as well as the authorities may experience demanding situations and critical moments which ask for an effective and quick response.

Although the study of disasters is in steady progress, there still remains a lot to be done. One of the blind spots in disaster research continues to be cross-national knowledge of disasters and disaster management. Certainly, during the last decade a number of encouraging efforts have been made to promote a more international perspective in disaster management. But as disasters and crises will increasingly feature transnational and even global characteristics, in that sense disaster and crisis research are lagging behind. Students of disasters and crises cannot limit themselves to typically domestic and local events. They should indeed cross the borders and engage in cross-national studies which fit in with the transnational dimensions of contemporary and future disasters and crises.

This study of flood response and crisis management in four Western European countries tries to appreciate that agenda. In a way, in this context it appears self-evident to examine disaster and crisis management in France, Belgium, the Netherlands, and Germany. In 1993 and 1995, all four countries suffered major inundations mostly caused by riverine floods which, of course, did not respect the administrative and political borders between them. A study of this kind of disaster must be cross-national. It must bring to the fore the similarities and differences in flood management between the countries involved. And it must solicit lessons for international cooperation in disaster warning and disaster response.

This opportunity for cross-national disaster research has been grasped by researchers from the four countries. They have bridged the usual differences in language – French, German and Dutch – and they have accommodated the ingrained differences in scholarly perspective which are so typical of the European academic scene. The final product reflects the sustained and painstaking efforts on the part of the international experts to provide a truly comparative perspective to the 1993 and 1995 flood response and crisis management in these four Western European countries.

For the years to come, it is hoped that this book may set the agenda for more comparative studies. The 1996-1997 floods in Central and Eastern Europe have revealed the vulnerability of the infastructures of the public works. It seems to be opportune to exchange expertise about disaster warning, response and relief between the various parts of Europe.

This study was sponsored by the Directorate-General for Environment, Nuclear Safety and Civil Protection of the European Commission; we have welcomed the encouragement and open-mindedness of the Directorate-General. The study has benefited considerably from the exchange of views at flood management seminars held in Florival (Belgium), Rome, Lisbon, and London in 1996 and 1997.

The coordinating group at the Crisis Research Center of Leiden University is grateful to its collaborators in France, Belgium, and Germany. The truly comparative feel of this book is largely due to their willingness to cross national borders.

Crisis Research Center
Leiden University

Uriel Rosenthal
Paul 't Hart

CONTENTS

List of figures and tables xi

Chapter 1

Flood response and disaster management: a comparative perspective

Uriel Rosenthal, Paul 't Hart and Michel Bezuyen

1 The 1993 and 1995 floods in Western Europe 1
2 Risk communication and warning 3
3 The organization of disaster response 4
4 Recovery and damage compensation 9
5 Study design 11
Notes 12

Chapter 2

Flood management in France

Claude Gilbert and Cécile Gouy

1 Introduction 15
2 Warning process and risk information 17
3 Disaster response 27
4 Recovery and damage compensation 40
5 Concluding observations 51
Notes 54

Chapter 3
Flood management in Belgium

Hugo van Hassel and Leen van Lindt

1 Introduction 57
2 The 1993 and 1995 riverine floods: an overview of events 59
3 Disaster management in Belgium 61
4 Risk communication 69
5 Disaster response 81
6 Recovery and damage compensation 92
Notes 99

Chapter 4
Flood management in the Netherlands

Uriel Rosenthal, Menno van Duin, Michel Bezuyen, Marie-Louise de Vreeze-Verhoef and Paul 't Hart

1 Introduction 103
2 The 1993 and 1995 riverine floods: an overview of events 104
3 Risk communication 113
4 Disaster response 129
5 Damage compensation 143
Notes 150

Chapter 5
Flood management in Germany

Wolf R. Dombrowsky and Lutz Ohlendieck

1 Introduction 153
2 The 1993 and 1995 riverine floods: an overview of events 154
3 Water management and disaster response in Germany 157
4 Risk communication 162
5 Disaster response 168
6 Recovery and compensation 182
Notes 186

Chapter 6

Flood response dynamics: Local resilience and administrative flexibility

Uriel Rosenthal, Michel Bezuyen, Menno van Duin and Paul 't Hart

1 Introduction: a constrained comparison 189
2 Risk communication: flood forecasting and warning 190
3 The organization and implementation of disaster response 193
4 Managing evacuations 201
5 The politics of damage compensation 207
6 Organizational learning 213
Notes 216

Appendix 1 Sources 219

Appendix 2 Floods in Western Europe 231

List of contributors 235

LIST OF FIGURES AND TABLES

Figures

2.1	The organization of disaster management in France	16
2.2	Organization of the warning process in France	19
3.1	The 1993 floods in Belgium	58
3.2	The 1995 floods in Belgium	60
3.3	The structure of disaster planning in Belgium	62
3.4	Emergency response during the 1993 and 1995 floods	82
4.1	The Dutch part of the Meuse river basin	107
4.2	The 1995 floods: stricken areas	109
4.3	The organization of emergency management at the local level	114
4.4	Intergovernmental relations during major disasters	115
5.1	Flood response organization in the city of Koblenz	170
5.2	Flood response organization in the city of Cologne at moderate water levels	176
5.3	Flood response organization in the city of Cologne at high water levels	178

Tables

3.1	The organization of emergency management in Belgium	65
4.1	Highest water level of the river Meuse in 1993	106
4.2	Highest water level of the river Meuse in 1995	112
4.3	Inter-regional time lags in warning and response start-up	119
4.4	Timing of warnings: citizens' evaluations	124
4.5	Primary source of information in Limburg 1993 and 1995	125
4.6	Primary source of information in Gelderland 1995	126
4.7	Quality of municipal information provision	127
6.1	Types of evacuations in 1993 and 1995	203
6.2	A typology of flood damages	208
6.3	Typology of compensation systems	210

Chapter 1

FLOOD RESPONSE AND DISASTER MANAGEMENT: A COMPARATIVE PERSPECTIVE

Uriel Rosenthal, Paul 't Hart and Michel Bezuyen

1 The 1993 and 1995 floods in Western Europe

During 1993 and 1995, several countries in Western Europe experienced major floods. Inundation of large cities and municipalities throughout France, Belgium, Germany and the Netherlands disrupted community life for several weeks, resulting in substantial damage to corporations, farms and households.[1] The floods demonstrated once again the continued vulnerability of modern societies to natural hazards. Although vast amounts of money have been spent on the management of river systems and coastlines, using more and more advanced techniques, the forces of nature still act in ways that both surprise and harm communities. For example, the floods of 1993 and 1995 were mostly provoked by enduring and heavy rainfall. More generally, however, the effectiveness of flood prevention measures has been partly offset by parallel developments increasing the susceptibility to floods. These include the creeping erosion of upstream mountain slopes and the sometimes extensive use of flood plain areas for housing and industry. Flood plain management in these countries is often a disjointed process where diffuse networks of authorities and community actors seek to prevent and mitigate flooding yet to some extent accept the safety trade-offs resulting from their pursuit of socio-economic development in high risk areas.[2] As a result, floods are most likely to remain a recurrent phenomenon along many of Western Europe's main rivers, placing a premium on effective disaster warning and response systems along their trajectory.

In this study, a four-country collaborative research team examines the operation of such systems under the pressure of extraordinarily high water levels. Floods are conceptualised here as disasters. Social scientists have tended to view disasters as extraordinary events generating collective stress and serious disruption of community life.[3] In recent studies it is emphasized that disasters are not necessarily life-threatening. The decisive factor is the disruptive impact of the pervasive uncertainty and widespread fear triggered by some occurrence or disturbance.[4] From the perspective of public administration, disasters are mostly taken to be situations that require a massive and coordinated deployment of organizational resources, often exceeding the capability of the local communities where the events are taking place. The 1993 and 1995 floods appear to meet both sociological and administrative disaster criteria.

From a disaster management point of view it is important to keep in mind that there are differences in the types of flood faced by the countries included in this study. The familiar distinction between disasters with a sudden and a gradual onset is relevant here.[5] The first type of floods studied are known as flash floods. These are characterized by an sudden and massive increase of water quantity leaving endangered communities only limited time to respond. Flash floods occurred in the southern part of France and will be discussed in detail chapter 2. In the other countries and in the northern part of France, disaster conditions were provoked by riverine floods marked by slowly rising water levels of main rivers and a gradual inundation of catchment areas. Flash floods impose a different and indeed more daunting challenge to the alertness of flood prone communities, whereas the larger scope and longer duration of riverine floods constitutes a mayor challenge to the scale and endurance of disaster management arrangements.

Taking the social disaster perspective as the main focus, the country studies are designed to highlight a number of classic crisis management challenges: the communication of risks and warnings of impending disaster threats; the organization of the response following the onset of disaster, including the pivotal decision making process about the evacuation of threatened populations; and the post-disaster mobilization of resources for recovery and compensation of individual and collective damages.[6] In the next sections of this chapter we shall briefly elaborate each topic, and emphasize the key questions driving the data collection and analysis by the participating researchers.

It should be emphasized at the outset that these themes by no means exhaust the range of questions one might ask about the 1993 and 1995 floods. In part, they reflect the preoccupations of researchers interested in public sector crisis management. Students of water management and flood response might well pursue different angles not covered in this study. For example, we do not describe or assess in any detail the infrastructural and other flood prevention schemes that existed prior to the floods, nor does this study go into any detail on the broader impact of the floods on regional and national water management policies. The lessons of this study, in contrast, are focused more on the coping with floods as social crises, and are hopefully of interest to emergency managers, disaster planners and public authorities dealing with a wide range of critical contingencies.

2 Risk communication and warning

In order to be able to take proper protective action against floods, people need to be well informed of flood hazards in general and acute threats of flooding in particular. Effective warning systems maximise the community's ability to adapt to the impending danger and hence to minimize the losses and trauma produced by the disaster. Disaster researchers generally agree that warnings need to be timely, accurate, reliable and credible to the receivers – be they government agencies, businesses or private citizens.[7] Warnings should contain factual information about the situation as well as clearcut advice on how to take protective action. Warning messages should be communicated via multiple channels. Ideally, warning systems produce a general, impersonal messages as well as more detailed information attuned to the specific situation of particular communities and target groups. Moreover, warnings are likely to be more effective when they are embedded in an integrated system of flood forecasting, warning and response. Comparative research suggests, however, that for many European countries such integrated systems are still far from being achieved.[8]

Flood warning begins with detection. Detection involves both the prediction of water levels and an analysis of their community consequences. To this end flood warning services may use a combination of meteorological and hydrological data. Particularly at the local level, less formal risk assessment methods relying on the intuition and rules-of-thumb of experienced flood watchers may complement if not substitute for official warnings by regional or national agencies when these are not forthcoming or appear to be at odds with local experiences.[9]

Following detection, warnings need to be disseminated. Messages need to be constructed to fit the information needs and frame of reference of different target groups. A well-known problem of disaster warnings is that those who produce warnings mistakenly believe that their messages are received by perceptive individuals and communities, and that they will take immediate action after having been informed. This is not the case. The dissemination and reception of disaster warnings is a social and not a mechanical process.[10] Three factors are important here: the clarity of the message, the credibility of the source, and the risk perceptions of people.

If warning messages are couched in technical language or contain too many 'ifs' and 'buts', this increases the scope for recipients to simply discount or misinterpret them. Also, official warnings do not necessarily have high credibility. In fact, people generally tend not to rely solely on official communications. They check local media for their reading of the situation, they talk to neighbours, they call friends and relatives. Official warning agencies often fail to take into account the role of these informal networks.[11] A special credibility problem relates to the so-called 'cry wolf syndrome': the boy that cried wolf at every strange sound in the forest was eventually no longer believed by his friends, even when he got it right.[12] Likewise, official agencies that have predicted and warned the public about floods that subsequently did not happen may lose credibility. Knowing this, they will want to protect their reputation and develop elaborate verification procedures, at the risk of losing valuable time and trading off speed for accuracy. Finally, the impact of warnings depends on people's assessment of the risks they are facing. Their subjective risk perceptions are shaped by many factors of which official flood warnings are only one. Prior experiences with high water are very important here, as is the feedback people obtain from others in their immediate environment.

Similarly, there are many complex social and psychological factors involved in people's decisions what to do once they have received disaster warnings that they regard as credible.[13] People weigh the risks of staying put and doing nothing against the presumed consequences of alternative courses of action. Citizens' decisions on how to respond to warnings are obviously not a matter of systematic cost-benefit analysis, nor are they the product of irrational fear and panic, however. The low incidence of panic in disaster situations is by now extensively documented, yet to this day many planning and warning agencies withhold relevant information or delay warnings because they fear the public will act irresponsibly.[14]

In the risk communication parts of the country studies we shall describe the warning process during the 1993 and 1995 floods. The case study authors have described the interorganizational warning and mobilisation process among the relevant public authorities as well as the dissemination of warnings from the authorities to the public. Specifically, they have briefed to examine the role of official and informal warning practices, the use of different types of information in the flood detection process and the use of mass media as channels for transmitting warnings to the population. Moreover, in the Dutch case study, survey research has been conducted by the authors to gain insight into the reception of warnings by the public, highlighting the use of different information channels and the differential warning time and message quality as experienced by receivers in different regions affected by flooding.

3 The organization of disaster response

The onset of a major flood sets in motion a large-scale and complex coping effort that involves political administrative leaders, rescue agencies and other public services. The official response runs parallel to the activities of voluntary organizations, mass media, community groups and individual citizens. With so many people and organizations mobilising and taking action, planning and coordination are important to make sure response efforts are successful. Two success criteria are particularly important: comprehensiveness and effectiveness.

Disaster response should be comprehensive in that it fully covers the stricken or endangered area and meets a broad range of urgent needs and problems arising in the community. Response effectiveness is a matter of providing speedy and professional emergency assistance to flood victims. This includes among others rescue, transport, shelter, medical care, material support, and psychosocial assistance to flood victims. The effectiveness of emergency operations hinges upon the quality of the response process as indicated by the provision of adequate intelligence, communications, coordination and logistics within and between the parties involved. Furthermore, effective disaster response is not a matter of officials and public agencies acting autonomously. Official responses can only be successful if they proceed on the basis of realistic expectations about the behavior of the population in the stricken area and if they take into account the self-organizing capabilities of local citizens.

Disaster response is an occasion for the exercise of leadership in making crucial decisions about the allocation of resources and treatment of victims and other parties to the disaster.[15] Furthermore, response leadership requires an astute awareness of the socio-emotional dimension of disaster, and the importance of material and symbolic efforts that can be made by authorities in order to help stricken communities cope with the collective stress generated by the events.

A key issue in organizing responses to major floods involves scale, e.g. finding an appropriate balance between the geographical and social scope of the disaster on the one hand, and the administrative division of labor in managing the response. Whilst some floods remain confined to sudden outbursts of high water in particular locations, many span tens if not hundreds of kilometres along the river banks. In that case, they are likely to transcend jurisdictional boundaries, both within and across countries. Responding to major floods requires more than just parallel efforts by local authorities and communities to deal with their segment of the problem. Many times local needs for personnel and resources to combat the effects of flooding exceed local capabilities, particularly those of smaller communities. Moreover, convergence of rescuers to disaster sites and the movement of large numbers of people and livestock away from them tends to produce major logistical and transportation problems. In short, major floods tend to induce an upward pressure in the organization of response efforts. Local communities need to coordinate their efforts, and often problems and needs arise that can only be met by the involvement of regional or national administrative bodies.

A tension exists here between established administrative doctrines that emphasizes the utility of centralization of crisis response coordination, and the emphasis placed by many studies of community disasters on the pivotal role of the local level.[16] Many official crisis plans not only call for a concentration of decision making authority in crisis centers under clear executive leadership. In the case of large-scale or highly consequential crises they also provide for administrative centralization, e.g., taking coordinating powers from the local level upward to regional and national crisis centers. This is held to enhance uniformity of the response, speed up the decision making process, and help overcome fragmentation as well as inter-local and interorganizational coordination problems. In contrast, empirical studies of crisis management draw attention to the value of decentralization and flexibility in disaster response. Local communities and operational commanders of rescue services are closest to the problems at hand. They can deal most quickly and effectively with peculiarities of local situations. From their point of view a centralized structure may constrain rather than strengthen speed and flexibility.

Different countries may deal with this problem in different ways. In the United States, for example, intergovernmental tensions are a conspicuous phenomenon in disaster management.[17] In the US, tensions center around the granting of presidential disaster declarations, the possible 'take-over' of the disaster by the Federal Emergency Management Agency, and the maze of overlapping and competing federal agencies with disaster management responsibilities.[18] In Europe, with its variety of administrative systems and in some cases stronger tradition of local government, we may expect differing degrees of centralization-decentralization and encounter other intergovernmental fault lines in disaster management.

To examine this issue, the administrative structure of disaster response as it evolved during the floods of 1993 and 1995 will be analyzed in the country studies. We shall describe the organization of disaster response with special emphasis on the role played by local, regional and national authorities. A 'critical event' in this regard is the decision making process regarding evacuation, both for individual citizens and public authorities.[19] From a public policy perspective, deciding upon the large-scale evacuation of communities stricken or endangered by high water levels poses one of the most poignant challenges in flood response. It involves difficult trade-offs between the economic costs and social disruption of such massive operations, and the often uncertain but potentially catastrophic consequences of not evacuating at all. Especially in the Dutch case, evacuation decision making provided the crucial test of the disaster response structure, and put into sharp perspective the problems of coordinating local, regional and national level perspectives and policy agendas.

When reading the country studies on the subject of disaster response it is useful to keep in mind the basic institutional make-up of the governmental system in France, Belgium, the Netherlands and Germany. A brief outline of these is provided here.

France — France is often, but only partially correctly, viewed as the prime example of a centralized unitary state. The national level, headed by the President, plays an important role in the administrative system of France. The French governmental system is further divided into a system of departments and communes called municipalities. The department is a crucial territorial administrative echelon in France. The 97 departments are headed by Prefects appointed by and answerable to the government in Paris. Elections for the departmental councils – *Conseil General* – are held on a cantonal basis: each canton (which may consist of a group of municipalities or even one part of a very large municipality) elects one member to serve for a period of six years. Departmental councils deal with the administration of state services, such as roads and schools. Departments in turn are clustered in 22 regions. In the department where the regional seat is located, the state is represented by the regional prefect. The regional prefect, however has no hierarchical power over the other prefects, but plays a coordinating role. The local level consists of municipalities *(communes)* each made up of an elected municipal council and mayor. Municipal councils are elected for six years, and the council chooses its own mayor from among its elected members. The mayor acts as the head of the local government and is subordinate to the Prefect who is the representative of central government. The mayor is responsible for the main functions of the municipality, which include public order and safety and other functions delegated by higher authorities.

Belgium — In recent decades, Belgium has gradually transformed from an unitary into a federal state with a complex and multi-layered administrative system. Five types of territorial authority structures co-exist: the federal government, the community, the region, the province and the municipality. The federal government performs key state tasks including national defence, law, public finance, social security and foreign policy. There are three communities in Belgium, structured along linguistic lines: Flemish (Dutch), Wallonian (French), and Germanic (German). Each community has its own elected parliament and government. They have tasks in the field of culture, language policy, education, sports and so forth. The regions are the relevant administrative entity when it comes to economic issues: the Flemish Region, the Brussels-Capital Region and the Walloon region. The regions are responsible for tasks such as regional development, housing, supervision of the provinces and communes and employment. At the same time, Belgium remains divided into provinces and municipalities. The provinces have directly elected councils, a permanent delegation for the daily administration and a governor who is appointed by the federal government. The province performs tasks such as maintaining law and order and the maintenance of roads. The province also exercises supervision over the municipalities. Belgium has 589 municipalities with an elected local district council, a council of aldermen and a mayor. The executive power rests with the council of aldermen presided by the mayor. Municipalities have tasks in the field of maintaining law and order, organizing elections and other social activities.

The Netherlands — The Netherlands is a decentralized unitary state with three levels of government, i.e. central government, twelve provinces and over 600 municipalities. An elaborate system of co-governance between local and central layers of public administration has developed, with the provincial level traditionally being relatively weak. Each municipality has an elected municipal council, a mayor and an executive Board of Mayor and Aldermen. The Board of Mayor and Aldermen, which often consists of full-time professional administrators, plays a leading role, while the Council generally limits itself to legislative oversight and control of the Board's policies. The Aldermen are elected from the Council, while the mayor is appointed by the central government. Historically, mayors were supposed to act as the eyes and ears of the central authorities, and to enforce central rule if necessary. Nowadays, mayors have become almost purely local officials who will not shy away from tough negotiations with their superiors at the Ministry of the Interior in The Hague to defend local interests. Municipalities have both autonomous and co-management tasks. The most important tasks delegated to or implemented by the local government are the maintenance of public order and fire protection, the upkeep of roads, and several tasks related to public transportation, economic policy, primary education, culture, social welfare and health care. In water management, an important role is played by so-called polder board districts. Founded in the Middle Ages, these are regional, self-governing administrative bodies of elected representatives and a permanent staff tasked with the management of water ways and water levels on their territory. At the national level, *Rijkswaterstaat*, a division of the ministry of Transport, and its provincial branch offices are in charge of the construction and maintenance major water infrastructures including sea and river dikes.

Germany — Germany is a federal state. The German Constitution distinguishes between the powers of the federal government in Bonn (over such matters as defence, foreign affairs, the national budget, social security and international trade) and the states – *Länder* – (such as local government, police and education). Each state has its own constitution, government and legislature. Major differences among their administrative systems make it difficult to compare the states. The central level of the state usually consists of ministries of State (e.g. ministry of the Interior, ministry of Environmental Affairs). The next level is the communal level which can be distinguished in regions, districts, administrative areas and local communities. Cities such as Bremen and Hamburg are simultaneously districts and units of local government, while other states comprise three levels: regions (for example the state of North Rhine-Westphalia has six regions), districts and local communities. Some communities work together in so-called administrative areas *(Kreise)*, while others do not need to cooperate *(Kreisfrei)*. Administrative fragmentation is also visible in local administration, where there are intricate divisions of responsibility between local councils, mayors and judicial authorities.

4 Recovery and damage compensation

The crisis atmosphere induced by a major flood does not subside when the immediate physical danger has disappeared. The post-disaster recovery period may in fact be turbulent, controversial, frustrating and even traumatic in its own right.[20] Evacuees may have returned to their homes, businesses may re-open their doors, and cleanup and repair work may have begun in local communities, the collective stress generated by the flood often has all but disappeared. To begin with, there may be significant numbers of victims and rescue workers who are physically harmed or suffering from post-traumatic stress. These require extended medical and psychosocial care. It is well known from disaster research that in disaster planning and management the 'first wave' of immediate emergency response receives far more attention than the organization of long-term care. The absence or insufficiency of a 'second wave' of psychosocial assistance may aggravate the plight of victims. It has proven exceedingly difficult in many cases to mobilise the necessary resources and political leverage for extended care programs.[21]

Secondly, there is likely to be controversy over the underlying and immediate causes of the disaster. At one level, floods may seem a classical example of a natural disaster. Indeed, without extensive rainfall, strong winds or in some cases unusually warm weather (speeding up the melting of ice in upstream mountain areas) there is usually no risk of flooding. Increasingly, however, in the public and political debate the occurrence of flooding is also connected to unsafe uses of flood plains for development and inadequate disaster prevention measures. Moreover, following many floods there is criticism of the warning process. A lack of timely warning has come to be viewed as a pivotal contributory cause of the scope and magnitude of disasters. Perceived failures of prevention and mitigation may highlight the man-made dimensions of flooding. They provide the focal point of intensive blaming and scapegoating in the aftermath of a disaster.

Thirdly, when the immediate threat is gone the organizations involved will start to investigate and evaluate the crisis. The evaluation process takes place at different levels and in different fora. In the public arena, political leaders and agency heads render account to legislative and oversight bodies, explaining and defending their policies and suggesting the lessons that might be learned. A similar but more detailed process goes on at the operational level, where internal reports and debriefing sessions provide the impetus for organizational learning initiatives.[22] When taking place against a backdrop of public criticism of pre-disaster policies and preparedness, disaster evaluations may become a tense affair. Victims, political opposition forces and other stakeholders may view the accountability process as an opportunity to punish or even oust officials, and to enforce hitherto unthinkable policy changes. This is mirrored by the mobilisation of organizational defenses and political survival instincts among incumbent authorities. Consequently, the motivation to learn from disaster experiences may be overshadowed by the forces of political expediency.

A particularly sensitive issue in the post-disaster phase concerns the compensation of victims and the allocation of resources to community recovery. In short: the question of 'who pays?' The salience of the issue is underscored by the experience of protracted court battles following major corporate tragedies, such as the Seveso tragedy, the Bhopal Union Carbide disaster and the Exxon Valdez oil spill. The underlying issue does not remain confined to industrial accidents and other man-made disasters, however. If floods and other natural disasters are becoming 'de-naturalized' as far as their causes are concerned, efforts to determine their causes become battles for liability whose outcomes may have enormous financial consequences for the parties involved. Moreover, if despite all efforts at decreasing their incidence, floods turn out to be a recurrent phenomenon, liability problems are not incidental but endemic.[23] To prevent time-consuming and divisive inquiries and litigation proceedings, countries may want to develop national or regional insurance systems spreading compensation and recovery costs. The issue involves both the compensation of different types of groups: private citizens, farmers and businesses, and public sector authorities and agencies. Likewise, the type of damages sustained may be differ. Some parties experience direct effects of water, others face revenue losses due, yet others incur major costs during the response operations. For each group and each type of damage different liability arguments may be constructed. In the country studies we shall describe how the recovery and compensation systems of the four countries were structured and how they operated in practice following the 1993 and 1995 floods. Relevant issues concern the legal status of insurance systems, the mix of public and private institutions involved, and the extent of coverage provided to various categories of claimants.

5 Study design

The floods of 1993 and 1995 provide a near-perfect laboratory experiment for students of crisis management. Covering the French, Belgian, Dutch and German experiences in both 1993 and 1995, this study allows for a two-way comparison. On the one hand, we can compare flood management within each country at two points in time, allowing us to observe the impact of one major flood experience on the system's ability to respond to the next. On the other hand, we can contrast the disaster management process in different countries responding to the same flood. In addition, in both cases we can also examine the process of transboundary cooperation between neighbouring countries in flood warning and response.

Taking up the challenge to exploit these opportunities for comparative study, the Crisis Research Center of Leiden University in the Netherlands put together a joint research group of international experts in the field of hazard and disaster management. The main objective of the project was to enhance the systematic understanding of national differences and similarities in flood management. A joint research protocol was developed to provide for a systematic and comparable reconstruction and analysis of the case events. Each case study is reported in a comparable fashion, highlighting the three core themes developed above. There are, however, some differences in the scope, methods and depth of the country studies, stemming from the methodological preferences, research resources and access to relevant data of the four teams.

Chapters 2 to 5 present the French, Belgian, Dutch and German experiences. In the concluding chapter, the national experiences will be compared and discussed in view of the disaster management literature. In the appendix, an overview of recent flood events in Europe is provided, as well as detailed accounts of the sources used.

Notes

1 Great Britain and Norway also experienced major flood problems in this period.

2 J. Handmer, (ed.), *Flood hazard management*, Norwich: Geo Books 1987; E. Penning-Rowsell, M. Fordham (eds), *Floods across Europe: Flood hazard assessment, modelling and management*, London: Middlesex University Press 1994; J. Handmer, Policy design and local attributes for flood hazard management, *Journal of Contingencies and Crisis Management*, 4, 1996, 189-197.

3 A.H. Barton, *Communities in disaster: A analysis of collective stress situations*, New York: Doubleday 1969; G. Kreps, Sociological inquiry and disaster research, *Annual Review of Sociology*, 10, 309-330. The term disaster is multifaceted, however, with different professions and scientific disciplines emphasizing different aspects. Meteorologists are mostly interested in the conditions which cause floods, physicians are interested in the injuries resulting from disasters, and so on. This study takes a social-scientific perspective.

4 K. Erikson, *A new species of trouble*, New York: Free Press 1994; E.L. Quarantelli, The future is not the past repeated: projecting disaster in the 21st century from current trends, *Journal of Contingencies and Crisis Management*, 4, 1996, 231-2.

5 Barton, *op. cit.*; a more elaborate typology of floods in Europe can be found in E. Penning-Rowsell, B. Peerbolte, Concepts, policies and research, in: Penning-Rowsell and Fordham (eds), *op. cit.*, pp. 12-16.

6 See, for example, R.R. Dynes, *Organized behavior in disaster*, Lexington: Heath 1970, chapter 5; R.W. Perry, *Comprehensive emergency management: Evacuating threatened populations*, Greenwich: JAI Press 1985; E.L. Quarantelli, Disaster crisis management: A summary of research findings, *Journal of Management Studies*, 25, 1988, 373-385; U. Rosenthal, P. 't Hart, M.T. Charles, Introduction: The world of crises and crisis management, in: U. Rosenthal, M.T. Charles, P. 't Hart (eds), *Coping with crises: The management of disasters, riots and terrorism*, Springfield: Charles Thomas, pp. 3-33.

7 H.D. Foster, *Disaster planning: The preservation of life and property*, New York: Springer Verlag 1980; Perry, *op. cit.*; T.E. Drabek, *Human systems responses to disaster: An inventory of sociological findings*, New York: Springer Verlag 1986,; E.C. Penning-Rowsell and J.W. Handmer (eds), *Hazards abd the communication of risk*, Aldershot: Gower 1990.

8 D. Parker, M. Fordham, J-P Torterotot, Real-time hazard management: Flood forecasting, warning and response, in: E. Penning-Rowsell, M. Fordham (eds), *Floods across Europe: Flood hazard assessment, modelling and management*, London: Middlesex University 1994 pp. 135-166; D. Parker, M. Fordham, S. Turnstall, A-M. Ketteridge, Flood warning systems under stress in the United Kingdom, *Disaster Prevention and Management*, 4, 1995, 32-42.

9 D.J. Parker, J.W. Handmer, The role of unofficial flood warning systems, Paper Flood Hazard Research Centre, Middlesex University 1996.

10 T.E. Drabek, *Human system responses to disaster: An inventory of sociological findings*, New York: Springer Verlag 1986, pp. 71-99.

11 Increasingly, however, the institutionalized use of low-treshold and widely trusted local radio and television stations for the transmission and interpretation of official disaster warnings is viewed as an effective way of bridging the gap between the technocratic world of forecasting and the social settings of target communities. See M.J. van Duin, M.J. Bezuyen, U. Rosenthal, *Evacuaties bij hoog water: Zelfredzaamheid en overheidszorg*, Leiden University: Crisis Research Center 1995.

12 E. Kam, *Surprise attack*, Cambridge: Harvard University Press 1988.
13 I.L. Janis, L. Mann, Emergency decision making: A theoretical analysis of response to disaster warnings, *Journal of Human Stress*, 3, 1997, 35-48; R.W. Perry, A.H. Muskatel, *Disaster management: Warning, response and community relocation*, Westport: Quorum Books 1984.
14 G. Kreps, Foundations and principles of emergency planning and management, in: D. Parker, J. Handmer (eds), *Hazard management and emergency planning: Perspectives on Britain*, London: James and James 1992, pp. 159-174.
15 U. Rosenthal, P. 't Hart, M. van Duin, A. Boin, M. Kroon, M. Otten, W. Overdijk, *Complexity in urban crisis management: Amsterdam's response to the Bijlmer air disaster*, London: James and James 1994, pp. 119-125.
16 See further P. 't Hart, U. Rosenthal, A. Kouzmin, Crisis decision making: The centralization thesis revisited, *Administration and Society*, 25, 1993, 12-45.
17 R.M. Scratton, *Disaster relief: The politics of intergovernmental relations*, Lanham: University Press of America 1989.
18 Scratton, *op. cit.*; A.K. Settle, Disaster assistance: securing presidential declarations, in: Sylves and Waugh (eds), *op. cit.*, pp. 33-56; W.L. Waugh, *Terrorism and emergency management: Policy and administration*, New York: Marcel Dekker 1990, pp. 32-38.
19 See, for example, A. Kirschenbaum, Warning and evacuation during a mass disaster: A multivariate decision making model, *Journal of Mass Emergencies and Disasters*, 10, 1992, 91-114.
20 See, for example, K.T. Erikson, *Everything in its path: Destruction of community in the Buffalo Creek flood*, New York: Simon and Schuster 1976; P.H. Friesema, J. Caporaso, G. Golstein, R. Lineberry, R. McCleary, *Aftermath: Communities after natural disasters*, Beverly Hills: Sage 1979; R. Geipel, *Disaster and reconstruction: The Friuli earthquakes of 1976*, London: Allen and Unwin 1982.
21 See P.E. Hodgkinson, M. Stewart, *Coping with catastrophe: Handbook of disaster management*, London: Routledge 1991, p. 77-103.
22 Such debriefings are also essential to enable first-line emergency personnel to 'let off steam' and begin to cope with their own traumatic experiences.
23 The costs of major floods are huge, and in various countries place the highest burden of all disasters on the national economy. Australia is a case in point, see M. Nakken, P.B. Mitchell, Reflections on flooding in central New South Wales: Learning from history and assessing the future, *Australian Journal of Emergency Management*, 11, 1996-97, 44-48.
24 Specifically for the Commission, a report version of the study was produced: U. Rosenthal, M.J. Bezuyen, M.J. van Duin and M.L. de Vreeze-Verhoef (eds), *The 1993 and 1995 Floods in Western Europe: A comparative study of disaster response in France, Belgium, the Netherlands and Germany*, Leiden University: Crisis Research Center 1997.

Chapter 2

FLOOD MANAGEMENT IN FRANCE

Claude Gilbert and Cécile Gouy

1 Introduction

France, unlike the other European countries included in this study, is exposed to both river and flash floods. Though the damage caused by both types of floods is considerable, flash floods capture more attention because of their suddenness, drama, and regularity in the south of France. Two particularly eventful flash floods occurred in Nîmes (department of Gard) in 1988 (9 deaths) and in Vaison-la-Romaine (department of Vaucluse) in September 1992 (41 deaths).

Between September 1993 and January 1994, 35 French departments were hit by either river or flash floods taking more than 30 lives and causing extensive damage. In January-February 1995, 31 departments were again affected, this time predominantly by river floods causing massive damage. Any analysis of the floods in France should take both types of flooding into consideration and should place particular floods in the context of all of the others which occurred in the same period. Three significant cases are reported here: the flash floods in Bollène and Valréas (department of Vaucluse) in 1993; the river floods in the department of Ardennes (around Charleville-Mézières) in 1993; the river floods in the Ardennes again in 1995.[1]

As a general background to the case studies, some preliminary information is needed about the French administrative system as it relates to disaster management. Power and responsibility for crisis management is shared between national entities and local communities. France is divided into 22 regions, and each region has an elected assembly *(Conseil Régional)* and a regional prefect who represents the state. The 22 regions are subdivided into 97 departments. The regional prefect also serves as the prefect of the department where the regional seat is located.[2] The departments have an elected assembly *(Conseil Général)* and a prefect who represents the state. Around 36,000 communes exist, and each has a municipal council *(Conseil municipal)* and a mayor who is a locally elected representative but who also represents the state at the municipal level. The department is the most important territorial administrative echelon in France, especially because a large number of the state's services are organized

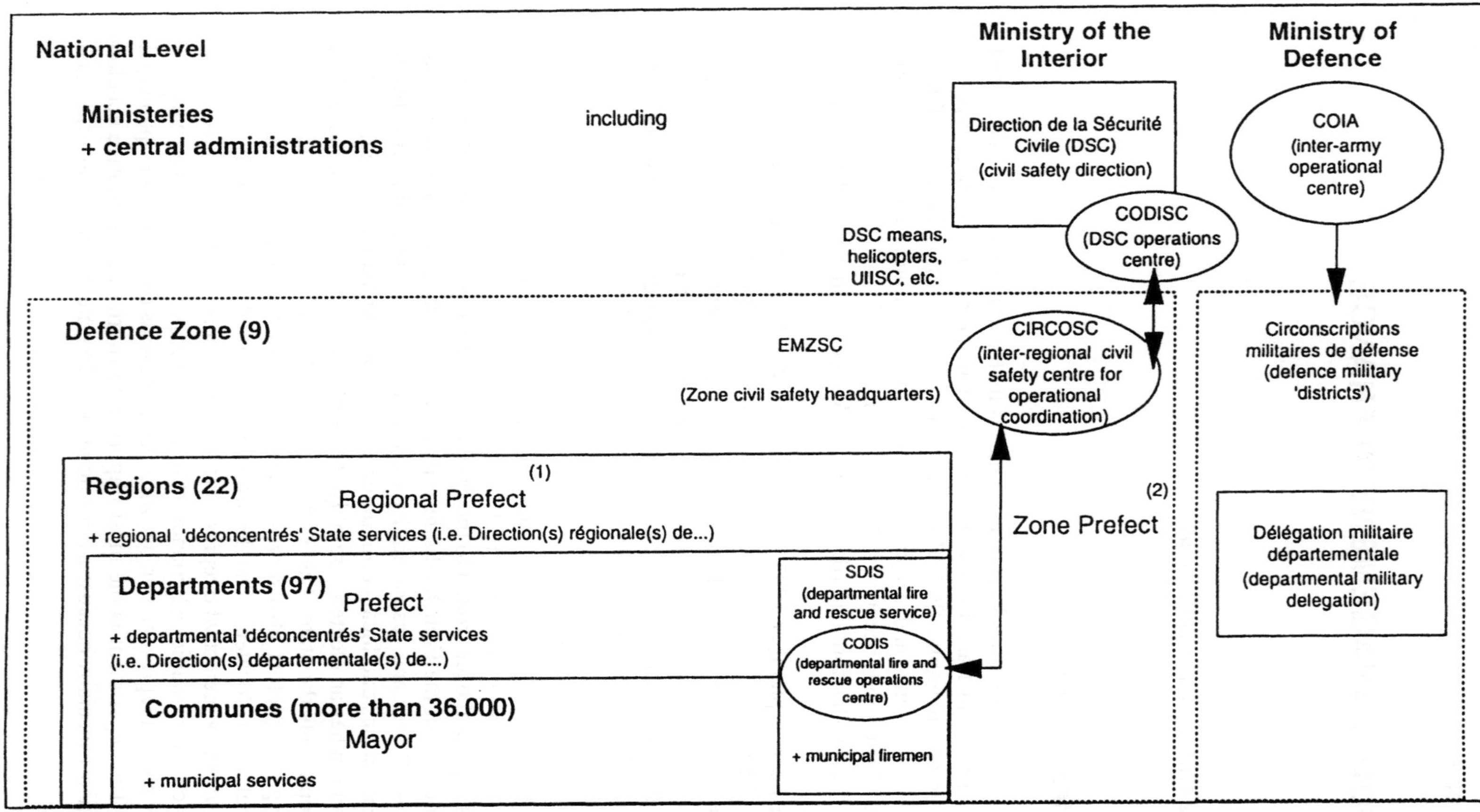

(1) The regional prefect is the prefect of the department where the seat of the Region is located.
(2) The Zone prefect is the prefect of the department where the seat of the Defence Zone is located.

Figure 2.1 The organization of disaster management in France

at this level.[3] Crisis management tends to be organized at the departmental level under the authority of the prefect.

In this chapter we shall describe the warning, response and recovery process in the 1993 and 1995 floods, focusing on the three cases for which material was collected.

2 Warning process and risk information

Flood warning in France

The flood warning system hinges on a distinction between the collection and processing of information on the one hand, and the alert itself on the other.[4] Information collection and analysis is done by two operational public services, the *Services d'annonce des crues* (SAC, flood announcement service) and the different meteorological services *(Météo-France)*. The alert is decided by the prefect and aimed at the mayors. Practically no formal procedures exist with regard to the transmission of information to the threatened populations. The mayors use all available means (radio, TV, newspapers and posters) to alert their citizens and communicate all relevant information concerning the imminence of danger and the actions to be taken.

The 52 *Services d'annonce des crues* (SAC services) keep watch over the waterways of the basin placed under their control (16,000 km out of 300,000 km of waterway are observed). During flood periods, their observations of river-levels and of rainfall are intensified. For this purpose, they have automated and manned observation stations (some 800 out of 1,200 are automated). River-levels are measured depending on the hydrological, climatic and topographical conditions of the basins. As for rain, the SACs use automated, tele-transmitting pluviometric stations. A few SACs have additional means allowing them to consult meteorological data (particularly radar images).

In case of high water levels the SAC prepares the alert and information bulletins for the prefect, who decides whether to pass them on to the mayors. Such alert bulletins are communicated via different services under the prefect's authority (such as the departmental fire and rescue service (SDIS), the national *Gendarmerie*, the national police, *France Telecom*). Once alerted, the mayors must take the initiative to provide updates by using a telephone message diffuser.

Each department has its own meteorological centre, the *Centre Départemental Météorologique* (CDM). In each region, these centres are under the authority of a *Centre Météorologique Inter-Régional* (CMIR). The interregional centres are, in turn, subordinate to a single body at the national level, the *Service Central d'Exploitation Météorologique* (SCEM) based in Toulouse. Together, these different structures form a national public body called *Météo-France*.[5] Only the CMIRs and the SCEM operate on a 24-hour basis.[6]

In addition to daily special meteorological communiques intended primarily for the media, these services send out bulletins warning against bad weather. The *Bulletin Météorologique Spécial* (BMS, special weather bulletin) announces storms, gales, snow and frosts and anything that might be considered dangerous. Since *Météo-France* produces several hundred BMSs a year, their warning function has been lost. Therefore, a special alert bulletin has been created: the BRAM *(Bulletin Régional d'Alerte Météorologique)*. The warning threshold has been determined jointly by *Météo-France* and the *Direction de la Sécurité Civile* (DSC, the civil safety directorate of the national Ministry of the Interior). The most urgent message is the A.L.A.R.M.E., an alert used only to warn for exceptional meteorological risks issued by the national SCEM in Toulouse. ALARMES are rare (about ten a year) and must clearly indicate the sort of bad weather (wind, rain or hail) and the areas at risk. The BRAM and ALARME are sent in a highly controlled manner. The ALARME is telexed to the operational centre of the *Direction de la Sécurité Civile* (CODISC) who transfer them to the relevant interregional coordination centres of the DCS (CIRCOSC); the BRAM are directly telexed to the CIRCOSC. These, in turn redistribute the bulletins to all the prefectures of the relevant departments To avoid mistakes, the telex is not altered in any way, and all of the prefectures are required to acknowledge receipt. Transfer time for the ALARME is supposed to be less than 15 minutes. It is up to the prefect to inform the mayors of the content of these bulletins upon reading them. The BRAM and the ALARME can also be sent to the media.

These structured and constrained formal provisions do not prevent informal, pragmatic warning arrangements from emerging. The actual warning system within departments or regions can vary considerably depending on local customs and particulars.

The warning process in the Vaucluse in 1993

The events considered here concern primarily the communes of Valréas and Bollène which are the largest, most urbanized and worst affected in the zone. Valréas and Bollène are located in the Lez river basin. Valréas lies upstream in the north of the department of Vaucluse, Bollène lies downstream in the western part of the Vaucluse.

The annual flow of the torrential river Lez is highly irregular. In the summer, the flow where it joins the Rhône is a mere $1m^3/s$, whereas it can exceed $13{,}000m^3/s$ during flood periods. Valréas, with a population of 9,000, is located in a hilly region and receives water from a tributary of the Lez, the Coronne, which itself is fed by various torrential streams (the Hérin, the Grand Vallat).

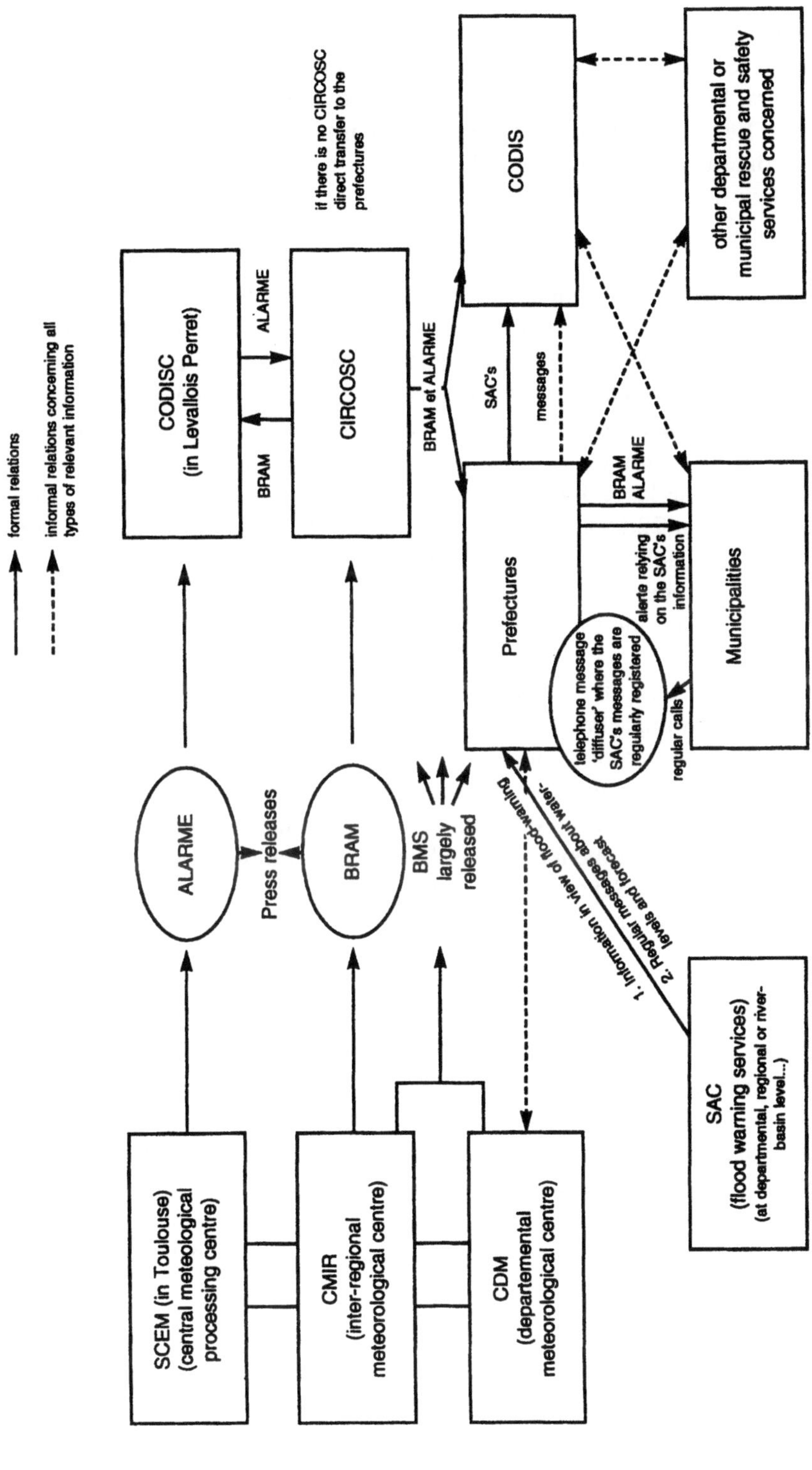

Figure 2.2 Organization of the warning process in France

The town is located where two torrents, neither of which can hold a substantial surplus of water, flow together and is therefore at risk from rapidly risingwaters. Bollène has a population of 13,000 and is located on the Rhône plain. The Lez flows through it and then joins a Rhône canal. Both towns belong to the same river basin, and if Valréas is struck by a flood, Bollène will be too a few hours later.

The causes of flooding — The southeastern part of France is regularly affected by violent storms in late Summer and early Autumn. Known as *cévennol* storms, they are caused by the condensation of cold air masses coming into contact with the still warm Mediterranean. Cévennol storms were particularly frequent in 1993. During the night of 30 September to 1 October, violent storms struck the north of the Vaucluse and the south of the Drôme. In Valréas, 162mm of rain fell in the evening of 30 September (a phenomenon which has only occurred twice in the last 30 years). This rainfall occurred between a first spell which lasted from 22 to 25 September flooding 16 southeastern departments (including the Vaucluse), and a third which affected the south and south east from 5 to 14 October. These rainy spells, notably those of 22-25 September, caused ground saturation and river obstruction.

The alert — The Lez, like most of the rivers in this part of France, does not have a flood warning service. Flood alert hinges above all on the information supplied by the meteorological services. For the period in question, neither BRAM nor ALARME warnings were issued. According to the departmental fire and rescue service based in Avignon several BMS (special weather bulletins) had reached the departmental fire and rescue operations centre (CODIS) in Avignon on 29 and 30 September, announcing the arrival of heavy precipitation and storms from the Mediterranean. However, later in the afternoon of 30 September, the bulletins no longer cited the Vaucluse among the target departments. The actual alert therefore largely came to rely upon local actors and particularly on the experience of the municipal firemen who give special attention to bad weather. In Valréas, as soon as heavy rain begins, firemen keep watch on the rising water level of the Lez, especially at a bridge which has a measuring scale. When the water rises above a certain level, they go on alert.

On 30 September, things happened very quickly. A first storm struck *Valréas* at around 20.00. The rain started to worry the firemen around 20.40-21.00 after they had begun to receive requests for assistance from individuals. They called in all off-duty personnel. The municipal technical services were informed at 21.00 and alerted their personnel. At 21.10, the local fire chief was requesting the departmental fire and rescue operations centre to send two response teams for reinforcement, while he went out on a reconnaissance mission. At 21.27, he prompted the mayor to go to the local emergency centre. At 21.37, having been told the mayor was on his way, the fire chief picked him up and took him for further reconnaissance. At 21.50, he called again for two reinforcement teams.

At 22.05, the mayor and the fire chief arrived in the district of la Gaillarde, an area of housing complexes which was getting flooded. They noticed that some people were already evacuating themselves and asked them to clear all other inhabitants from the ground and first floors. The fire chief called for reinforcements to look after them and left with the mayor for a nearby campsite, where thirty-some tourists and grape-pickers were staying. Though the Coronne had not yet flooded, they ordered evacuation of the campsites, being especially sensitive to the risks here. Special attention has been paid to campsites (at the local level as well as through national directives) since the 1992 Vaison-la-Romaine floods which had caused a lot of casualties among campers. A couple of Dutch tourists refused to evacuate (later, at 2.00, the husband, who had managed to cling to his caravan, reported that his 70-year-old wife had been swept away. The other death recorded in Valréas occurred at 22.46 when a 51-year-old wine producer died after a wall of his home collapsed when he attempted to break a hole in it to let the water flow out. The neighbour helping him had his legs crushed and had to be evacuated immediately).

Just then the town was plunged into darkness as the electrical transformer was submerged. Very quickly the streets turned into torrents of mud pouring into houses and in eight housing estates. Warning time was practically nil. With the water rising so quickly, it was too late to send a vehicle equipped with loudspeakers into the streets by the time the first official warnings came through.

At about 22.15, the mayor of the downstream town of *Bollène* – who had been alerted by Valréas counterparts and knew he had 5 or 6 hours before the water would reach his town – asked the municipal services to go on alert. The fire service, informed by their Valreas colleagues and the CODIS in Avignon, had already done so. Indeed, Bollène turned out to be far less overtaken by events than Valréas. The time lag was an important asset. It was also a case of prior planing paying off. A municipal safety committee had been set up in early 1993 composed of municipal leaders and relevant agency heads. In normal conditions, this committee conceives and prepares anticipatory measures. In times of crisis, it goes into action, as it eventually did on 1 October at 00.50. Way before that, however, its prime members and fire fighters had already taken precautionary steps.

As of 22.15, the municipal rain-collecting pipe network – which runs into the Lez – was closed to stop the Lez, already dangerously high, from rising in the sewers. From 23.00 onwards, first announcements were made by the fire service from a vehicle equipped with loudspeakers, aimed primarily at the most exposed inhabitants near the banks of the Lez. Around 23.48, the imminence of the flood was announced by telephone to the local campsite as well as to districts whose inhabitants were at particular risk from flooding. At 23.55, a similar announcement was made with loudspeakers in the three main streets of the town centre. The messages essentially communicated information and asked people to be vigilant and careful. They suggested evacuating ground floors, taking furniture upstairs, moving vehicles from the town centre and taking them to the two guarded car parks built on high ground (this request was made increasingly imperative as time went by). The messages also provided information about accommodation centres (a gymnasium and a secondary school prepared to this effect), their locations (it has been planned that the town would be split in two by the Lez, so there were centres on both sides) and advice on how to get there.

Around 1.00-2.00, the messages regarding evacuation and moving to the accommodation centres became more urgent. At 1.00, a second alert was made by loudspeakers and door-to-door visits in the most vulnerable districts. At 1.41, the siren was sounded repeatedly with five different modulations so everybody would understand the situation was urgent. The decision to use siren warnings was taken after firemen who had been sent to watch various sites and bridges upstream had communicated the rapid rise in water levels.

Despite these rapid and numerous precautions and attempts to inform the public, community reactions to the alert were sometimes sluggish. Many people seemed not to hear messages sounded at night. After the water had overrun the districts, the officials went around systematically banging on ground floor shutters. They were still waking people who were surprised to find their homes flooded and therefore had less time to react. The mayor and the fire fighters were also annoyed by townspeople coming out to merely watch the water instead of taking precautions to protect their own homes. Some rescue workers found themselves having to argue with spectators, receiving insults and having to force the on-lookers to leave. The removal of cars also proved to be a long and difficult operation which was not carried out in the best of circumstances. As a result, when the streets were flooded, some vehicles floated around chaotically and hindered the rescue operations of the fire department.

The warning process in the Ardennes in 1993

The Meuse, which rises in the Haute-Marne to an altitude of 384m and exits French territory in Givet in the Ardennes, can be divided into four sections. The high basin (the departments of Vosges and of Haute-Marne) is a sloping region of predominantly impermeable ground, which explains the rapid run-off of rainwater and of melting snow. In the department of Meuse, the basin narrows into a long straight stretch where the flatter and more permeable ground allows the water to filter through, thus diminishing the flow of the river. The basin opens out again in the department of Ardennes, where the ground is again impermeable, and the tributaries join the Meuse causing it to swell. The river narrows in the stretch between Charleville-Mézières and Givet and becomes a mountain gorge with a limited overflow area. Though all of the territories bordering the Meuse are affected by flooding, this study will limit itself to the situation in the department of Ardennes, and particularly to the communes located between Givet and Charleville-Mézières. Charleville-Mézières, which has a population of 60,000, is the seat of the prefecture.

Flooding of the Meuse is relatively frequent. It is caused by overflows of both the river and its Ardennes torrential-type tributaries. These floods generally evolve slowly and predictably, and they allow rescue workers to carry out preventative evacuations. In the rather agrarian department of Ardennes, the Meuse flows along several communities, both small and medium-sized, which have developed significant and diversified industrial activities. Charleville-Mézières is particularly at risk since the Meuse runs through it at three points (hence the large number of bridges linking the various districts).

The causes of flooding — An accumulation of rainy spells since the beginning of December 1993 (with rainfall two or three times greater than normal) and a warm spell in the Vosges mountains which caused snow to melt, took the rivers of the Rhine-Meuse, Artois-Picardie and Seine-Normandie river basins beyond their warning levels in mid-December. At first, 11 departments were affected, including the Ardennes. Human factors aggravated these natural causes, such as (often old) construction in low areas, poor maintenance of the banks and changes in agricultural practices which increase run-off.

The flood had two phases: an initial rise in water level in mid December followed by a lull; a further rise starting on 18 December which peaked on 22 and 26 December. In Charleville-Mézières, where the pre-alert level is 3m, in just 4 days the river rose to 5.78m on 23 December at 16.00. The level of the Meuse stabilized for a few hours before starting to drop regaining a less alarming, yet still high, level on 29 December (4.88m at 14.00). This flood moved upstream because of blockage in the tributaries downstream which is rather unusual and makes forecasting more difficult.

The alert — The SAC (flood warning service) observed the rising water levels with automated tele-measuring devices fixed in several places along the Meuse. After observing the rise in water levels, and receiving confirmation from meteorological services that the rain would persist, the SAC in charge of the Ardennes went into a 'state of vigilance' on 18 December.[7] It was mainly via

SAC data that the event was detected. The meteorological services had not released any bulletins since no severe rainstorms were forecast or occurring.

On 20 December 1993, the warning treshold was exceeded at several points in the Ardennes, particularly in Charleville-Mézières (4.12m at 8.00). The SAC warned the prefect that 'a state of serious flooding' had been reached and perhaps exceeded, and the prefect immediately set up a crisis committee in the prefecture. In the late evening of 20 December, the prefect warned the mayors in all of the threatened communes about the emerging threat and gave them the forecast of a further 30 to 40 cm rise during that night. The actual rise turned out to have been much greater, and some mayors later complained about the unreliable forecasts. The information from the SAC was updated frequently – five times a day at the peak of the crisis – and sent to the prefectures. The mayors must consult this information recorded on a telephone message diffuser once they have been informed by the prefecture.

The mayors, who know their areas quite well and are often familiar with recurrent flooding, did not wait for the official alert to get organized. The Charleville-Mézières municipality had mobilized its own observers. As soon as the river reached its overflow level (4m) a crisis committee was formed and three hundred people dropped their usual occupations and dealt fulltime with the management of the flood.

Throughout the flood, dissatisfaction existed among local elected representatives, who found the SAC messages unclear and imprecise since they gave readings for only 8 locations within the department. Moreover, the mayors and municipal services heads complained that the levels given were not immediately comprehensible: they would have liked the water levels to be expressed as NGF (*Nivellement Général de la France*, general French levelling) to enable the services to work from the national geographic institute's standardized maps. Instead the readings were expressed in terms of 'water levels of the rivers.' The SAC forecasts were also criticized by mayors who regularly noticed significant discrepancies between the forecasts and the actual situation in their parts of the river. To overcome some of these deficiencies, municipal staff and local elected representatives took the initiative in telephoning the prefecture and municipalities further upstream to learn about their situation in order to anticipate the rising water levels more accurately.

Around 19 and 20 December, steps were taken in various communes to raise furniture or to take it upstairs. In Charleville, before the town was really flooded, a free phone number was established to provide the public with official information from the municipality. Fluorescent posters were put up twice daily in risk areas showing the latest readings, a forecast and giving a few recommendations (such as: limit telephone calls, avoid using your car, remove your car from the area, find accommodation elsewhere). In addition, vehicles equipped with loudspeakers delivered messages in vulnerable areas. The prefecture used local radio stations to ask people from other places not to come to Charleville-Mézières. The departmental and municipal fire departments and other services gave technical (helping with breeze blocks and moving furniture) as well as psychological assistance (their presence alone was reassuring) to the community.

Generally speaking, many people were to some extent familiar with floods and the population reacted quite rapidly. Nevertheless, people did not always give sufficient credence to the threat announcements, claiming they knew what the risks were. In some communes many people, especially the elderly, refused to evacuate, preferring instead to keep watch over their homes and feeling adequately protected on their upper floors. Such refusals forced the rescue workers to make rounds by car or boat to try to convince people to leave, or if this failed, to supply them with food and check on their health. The only death recorded among the population (on 22 December) was that of a woman in her sixties in Givet who refused to evacuate and drowned in a metre of water inside her home after a fall and a dizzy spell. The other deaths were those of two Attigny gendarmes who disappeared on 23 December while patrolling the flooded areas by car.

The warning process in the Ardennes in 1995

The causes of the flooding — Rainfall during the last ten days of January 1995 in the Ardennes was two or three times greater than the average: 300mm of water from 20 to 29 January in the north of the department and 219mm near Charleville-Mézières, while the average for January is 100mm. Over the weekend of 22 January, 60 to 80mm were recorded. This relentless rainfall caused a rapid rise in water levels. It is important to note that because of their extent, the 1993 floods had been considered by all as a 'once in a century flood'. They were viewed as an exceptional event unlikely to repeat itself or to be worse only two years later. The situation in 1995 was one of people becoming somewhat accustomed to floods and believing the worst has come, but also one with a widespread feeling of astonishment and a loss of perspective in the face of an apparently endless rise in water levels. Indeed, in Charleville the flood lasted from 22-23 January to 8 February, with a peak on 31 January of 6.3m (52cm higher than the highest recorded level in 1993).

The alert — A number of lessons had been drawn from the 1993 floods. A new prefect had taken office shortly after the floods. It had been agreed to set up a more comprehensible warning system and to work on improving the forecasts. This initiative had been reinforced by the efforts of a national Parliamentary Inquiry Commission on the floods (with interviews in the Ardennes in June 1994), and by organizational reforms in the risk management field. The SAC (flood announcement service) was reorganized and placed under the sole authority of the Regional Environmental Directorate of Lorraine which covers the entire Meuse basin.

Since 1 December 1994, the SAC also possessed automated flood data collection and announcement systems. The new device made it possible to observe the flow of rivers in real time. Twenty-one stations and two radio relay posts were located in the Meuse basin. Data was collected every four hours during normal times and every hour in times of crisis. Afterwards, it was processed by computer to provide a two-hour forecast. When a 'state of serious flooding' would be declared, the SAC would communicate data every hour. The collection would be backed up by human observers. The SAC would also provide statistical progress charts to the administrative services, the municipal services, the *Conseil Général*, the media and the governor of the Namur Province in Belgium. Links with the meteorological services were also strengthened following the 1993 floods. The inter-regional meteorological centres in Lille and Strasbourg agreed to send a warning message to the SAC as soon as heavy rain is forecast and observed.

In 1995 the alert was characterized by the implementation of this new formal technical system and augmented by the informal adjustments made by the many actors involved. On Sunday 22 January 1995, the Strasbourg meteorological centre sent out a BRAM at 12.00 announcing heavy persistent rain, wind and snow melt-off. That day, as they observed the flow of the Meuse, various members of the technical municipal services in Charleville-Mézières contacted each other and sought information from the prefecture. The following day, the SAC told the prefect that in Charleville-Mézières, the Meuse had exceeded its overflow level (4m). The prefecture alerted the various basins and informed the mayors of communes bordering the Meuse and its tributaries. At the Charleville's town hall, the municipal staff went into a state of alert and began activities such as building walkways in the afternoon of 23 January, supplying breeze blocks and switching off certain electrical transformers. A committee directed by the mayor was set up to maintain dialogue between all the organizations involved: traffic, transportation, accommodation, safety, electricity, telephone and rubbish collection. Other communes also organized their efforts. The Charleville prefecture placed all of the operational services on alert and a prefectoral crisis committee was set up on 25 January.

The new system proved to operate relatively satisfactorily, and the interventions of additional observers from the North-Eastern Navigation Service in Charleville-Mézières and from the Gendarmerie in Givet were much appreciated since they provided an additional source of information. Throughout the crisis, the information supplied by the SAC via a telephone message diffuser was consulted by the mayors. Flood announcement charts were regularly communicated, especially through the media. There was less criticism on the part of local elected representatives than in 1993. Moreover, from the beginning of the flood and throughout subsequent events, Météo-France's weather bulletins played a significant role given the substantial precipitation. It would appear that, unlike in 1993, there was more time available for implementing preventative and anticipatory measures

As in 1993, the town and the prefecture regularly informed the inhabitants and asked them to comply with their recommendations. Information was posted on fluorescent municipal posters – by then a familiar sight for the population – and communicated through vehicles equipped with loudspeakers, the town's electronic notice boards, press releases and the local media. A free phone number was quickly established by the municipality to communicate with the public about issues such as re-housing, food supplies and counselling. As in 1993, official public messages gave updated forecasts and outlined the necessary precautions regarding property and safety. The prefecture issued requests to the public to limit their movements, for their own safety as well as to reduce damage to roads and bridges. In both 1995 and 1993, the local media largely acted as a conveyer of information from the authorities, yet did not play a major role in the warning process.

The local community responded better than in 1993. With the benefit of recent experience (the 1993 floods), the public more readily accepted the advice of rescue workers. Precautions were taken more calmly and less urgently, especially during the first few days. Nevertheless, refusals to evacuate again created problems for rescue workers even up to the point of compulsory evacuation being considered. Paradoxically, since the 1993 flood was held to be a 'once in a century flood', neither the public nor the rescue workers thought that the 1995 events would be very serious. They felt a repetition of the 1993 safety measures would do. As the days went by, however, the situation turned out to be more severe.

3 Disaster response

The organization of flood rescue operations in France

In France, according to the terms of article 1 of the law of the 22 July 1987, 'the purpose of the *sécurité civile* (civil safety) is to prevent all kinds of risks and to protect persons, property and the environment from accidents, disasters and catastrophes.' This same law broadly defines the organization of crisis management in case of civil safety crises which include floods. Formally speaking, mayors still have a primary responsibility to 'carry out necessary rescue work in cases of *fléaux calamiteux* (calamity).'[8] As a result of a steady evolution over the last few decades and of numerous legal provisions (especially the 1987 law), the management and implementation of public and private rescue operations come under the responsibility of the prefects when emergency plans are put into effect. The part played by state actors and institutions at the national, regional and *Zone de Défense* (Defence Zone) level is equally important, particularly those that come under the Ministry of the Interior. As far as floods are concerned, the deployment of means belonging to the state or placed under its authority is significant, since the personnel and equipment mobilized belong primarily to the departmental fire and rescue services (SDIS), the Civil Security Directorate of the Ministery of the Interior (DSC) and the Ministry of Defence.[9] Generally speaking, the formal provisions leave the various state and local actors room to manoeuvre, and this favours dynamics between the actions of state services and authorities and those of local elected representatives and territorial collective services (primarily communal ones). The various formal systems and process are defined below with emphasis placed on the measures relating to the administration's actors in as far as they apply to the whole territory.

Response organization: the departmental level — According to article 9 of the law of 22 July 1987, the prefect is in charge of 'implementing public and private rescue means and, when circumstances necessitate it, can trigger the departmental *ORSEC plan (Organization des SECours*, organization of rescue).' In this case the command of rescue operations is placed under his authority. Moreover, whether or not the emergency plan is activated, the prefect has the right to requisition private means throughout the department.

The ORSEC plan, which is pre-established by the prefect, is a general measure to ensure the safety of the population, particularly when the scope of a disaster transcends the boundaries of a single commune and seriously affects the lives of a large proportion of the population and the public services. The ORSEC plan is a generic emergency plan which is also to be used in case of floods. It defines the roles of the various services, and the terms of their collaboration. It allows rescue means from outside a department to be mobilized and coordinated. However, it is not implemented frequently or systematically since reinforcements from outside a department can also be mobilized informally.

With or without the ORSEC plan, the general idea encouraged by the Ministry of the Interior is to bring about a power increase in the decision making system.[10] This is achieved mainly through the establishment of a crisis committee at the prefecture as soon as a serious event is announced. The committee is placed under the authority of a member of the prefectoral corps. It brings together representatives from the departmental fire and recue services, the national police (Ministry of the Interior), the national Gendarmerie (Ministry of Defence), as well as representatives from a number of deconcentrated state services, technical experts and, possibly, the heads of the municipalities concerned. The prefect is supposed to coordinate the actions of the different state services at the departmental level and to maintain tight links with the heads of local services affected by the risk.

From a more technical point of view, the power increase of the crisis management system expresses itself through the creation of two command posts: a fixed command post and an operational command post (PCO). The fixed command post is located in the prefecture's operations room and has a two-fold function: to create a staff headquarters structure for the crisis committee and to work out the likely evolution of the crisis and its consequences. It should seek information, support the personnel involved, synthesize, inform and communicate. The PCO is where the tactical command of the operations is carried out. Its role is to conduct the operations, to coordinate the actions of the various services involved and to address requests for further means to the fixed command post. The PCO is under the authority of the prefect if several services are involved. It is often the director of the departmental fire and rescue services (SDIS), who runs the operation and coordinates operational activities.

Firemen, both professional and volunteer, play a decisive role in flood response. In France there are approximately 200,000 volunteer firemen (86% of the total) and 31,500 professionals including 8,500 military firemen (in Paris and Marseille). The fire departments are organized in operational structures of varying size, and with varying proportions of professionals and volunteers. Several types of structures can be distinguished depending on the territorial level, the number of interventions and the human and material means that they possess: *Centres de Secours Principaux* (CSP) and *Centres de Secours* (CS) for large towns or several communes and *Centres de Première Intervention* (CPI) for small communes, and possibly inter-communal centres.

The departmental fire and rescue service (SDIS) is a public body which in time of crisis directs the material, human and financial means of all of the fire departments in the department (municipal fire brigades, fire brigades of municipal public bodies and firemen of the SDIS). The SDIS usually has an operations centre (CODIS) which operates non-stop and in some departments, centralizes all emergency calls made by dialing 18.[11] It is the body in charge of coordinating the operational activity of all fire and rescue services within a department. It is immediately informed of all operations in progress, and receives regular bulletins regarding the evolution of the situation until the operations are terminated. The CODIS is placed under the authority of the departmental director of the SDIS and is responsible for liaising with the prefects, the departmental and municipal authorities and the other public and private entities participating in the rescue operation.[12] This integrated and hierarchical system does not exclude the existence of structures at the municipal level. Municipal crisis structures (such as crisis committees, command post, and emergency plans) can be organized by municipal representatives. Similarly, the municipal technical services and municipal fire departments can handle a large part of the operational crisis management.

Response organization: the Defence Zone national levels — The organization of the rescue operations at the departmental level is complemented at Defence Zone level. This territorial subdivision was established by law, in preference to the region, as an indispensable level between the state and the departments in terms of civil safety. The state representative in the department where the Defence Zone headquarters is located (also known as the zone prefect) prepares the safety measures and coordinates the public rescue services in the Defence Zone. In times of crisis, the zone prefect has the right to ascertain, allocate, centralize and prioritize the requests for all types of means (reinforcement personnel, equipment, resources) coming from departmental and regional prefects. He should call upon the military authorities for heavy equipment particularly suited to certain situations such as movable bridges, and lifting equipment. An interregional operational coordination centre (CIRCOSC) coordinates all rescue operations which demand the use of the zone public means or national reinforcements.[13]

At the national level the Civil Security Directorate of the Ministry of the Interior (DSC) is a major player in crisis management, notably its operations centre (the so called CODISC). This unit receives, completes and uses all available information concerning any event of interest to French civil safety as well as that of foreign countries. It reports to the cabinet of the Ministry of the Interior, responds to all requests for technical information coming from the prefectoral authorities and seeks and makes available to the local authorities all necessary rescue means and insures their distribution and coordination. On a national level, it liaises with the military operations centre of the army headquarters. The DSC also has its own equipment (helicopters, sanitary equipment) and staff ready for intervention. In total, it comprises of nearly 3,200 staff, both civilian and military (around 400 in central administration and 2,800 in operational services spread across the country).[14] Specifically, it has five sizable crisis intervention units (UIISC, largely made up of conscripts – some of them with 600 men) ready to respond throughout the national territory and abroad when a catastrophe occurs. DSC can send a partial or entire headquarters staff to support prefects who request it. DSC staff and equipment can be mobilized rapidly (especially via the Defence Zones).

Finally the organization of rescue operations involves a certain number of secondary tasks and mobilizes other types of actors who, for the most part, intervene to reestablish the major functions necessary for the day to day operation of collective services (including housing, food supplies, power networks, telecommunications, and water supply). These actors might belong to the civil service, local collective services (especially municipal ones), public and parapublic organizations, to private companies (sometimes requisitioned), or to local and national voluntary associations.

Crisis management in the Vaucluse in 1993

In both Valréas and Bollène, disaster management was organized primarily at the municipal level under the leadership of the mayors and with the participation and coordination of the departmental, zonal or national public operational services and the municipal technical services. The prefecture maintained its crisis committee which had been set up for the floods in Pertuis on 22 September 1993. This committee consisted of the heads of all the state services involved and met twice daily at the prefecture for briefing sessions. The ORSEC emergency plan was not implemented, since the prefect deemed it unnecessary. The response at the community level proceeded in a coordinated fashion and the operational means provided from the outside were equivalent to those the department could have obtained under the plan: 225 soldiers from the DSC intervention units and 220 firemen from around the region. Similarly, a hundred or so gendarmes and over 150 soldiers were mobilized, even several days after the onset of the flood. They participated in clean-up and surveillance operations. Staff from the Red Cross and other voluntary associations also played an active support role.

There were many experienced officials involved. The prefect – who was formerly the director of the Direction de la Sécurité Civile (DSC) – monitored the situation closely, via the crisis committee in particular, provided the support requested, organized communication with the media and visited sites late in the morning of Friday 1 October to assess the situation. He also accompanied the Minister of the Interior who visited Valréas and Bollène by helicopter late in the afternoon of 1 October to assess the extent of the damage and assure the victims of the government's sympathy and support. In addition to the personality and experience of the prefect, some past experiences of the main operational actors help to explain the specific precautionary measures, organizational choices and rescue actions taken. For example, the mayor of Valréas was also a policy executive in the Prime Minister's office and, in this capacity, had already been in charge of assessing crisis management during the 1992 Vaison-la-Romaine floods. Similarly, the Valréas fire chief had just arrived in 1993 after serving in Vaison-la-Romaine for a period of time including the 1992 September floods. More generally speaking, the mayor of Bollène and various staff members of both municipalities' services had been more or less involved in crisis and post-crisis management in Vaison-la-Romaine. These experiences seem to have played a formative role not only during post-1993 evaluations of pre-disaster preparation measures, but also in generating awareness of the need for strong personal commitment among the authorities involved.

Valréas — In Valréas on Thursday 30 September, 23.00, a crisis committee, a command post (PC), and various special subcommittees were set up. Under the authority of the mayor and fire service executives, the crisis committee consisted of officials from the fire department, the municipal technical services, the national Gendarmerie, the public utilities and various state services, particularly from the public works department. DSC intervention units arrived in the middle of the night from Pertuis where they had been stationed since the floods of 22 September 1993.

Throughout the night of 30 September to 1 October, various bridge-marking operations and evacuations were carried out. Two-hundred people were evacuated and lodged in a play centre located on higher ground. Following a request for assistance, the air force delivered 200 beds and sleeping bags at around 4.00. By 6.45, only 50 people were actually lodged in this centre due to a fall in the water level and many people being offered accomodation by local inhabitants.

In the middle of the night, all roads were closed but for two departmental ones, one of which was cleared continuously by public works vehicles to maintain access. At 3.00, EDF *(Electricité de France)* switched the power, which had been off since 22.45, back on. A search to find the body of the missing Dutch tourist began, using divers and boats. At 8.00 the Coronne was still flooding, and at mid-morning (10.30), violent rains began again and a torrential stream, the Grand Vallat which flows into the Coronne, devastated one housing estate. Some of its inhabitants who had not wanted to leave during the night and whose houses could no longer be reached, were taken to safety by DSC helicopter.

Most bridges were seriously damaged and the roads were furrowed making driving impossible. The public works department public works (or since) set up detours the same day. Department stores and the technical secondary school were closed due to an influx of mud. Many companies and farms were stricken. The water purification plant was completely destroyed, creating a drinking water problem. The health and social services department delivered two water tankers of 30,000 litres each and a truck loaded with bottled spring water. With its permission, the Red Cross established a water treatment and packing unit with an output of 1500 1.5 litre sachets/hour for the commune of Grillon. Apart from

Grillon, the drinking water supply for the entire Valréas/Bollène area was cut off. Three DSC mobile water purification units were delivered there on Saturday 2 October. Another mobile unit provided water for the hospital and the elderly citizens' home. That same day, Docters Without Borders distributed chlorine tablets to disinfect the water that was available. EDF's and France Telecom's worked frantically to restore power and telephone services throughout the area. The Red Cross distributed clothing and the catholic solidarity association Secours Catholique supplied the victims with clean-up equipment (shovels, scrapers). While these actions were being carried out, the Gendarmerie patrolled the area to prevent theft and looting.

Bollène — In Bollène, the flood was managed by the municipal safety committee using its Safety Plan drawn up at the beginning of 1993. The town was divided into intervention sectors, since the Lez cut the town in two. The organization of the crisis command was carried out under the leadership of the mayor. As of mid-evening, the municipal technical services and the fire department were reinforcing dikes, pumping water and informing the population before gradually evacuating the inhabitants. Accommodation centres (a gymnasium and a school) had been established, one on each bank of the Lez, in anticipation of the division of the town.

At 4.00, the Lez burst its banks and flooded the town centre and the town hall. Bollène was divided in half over a distance of 5 km and plunged into darkness after an electrical transformer was flooded. The situation was dramatic: through the darkness came the roar of the water, the shouts of the victims and the barking of dogs. The firemen carried out the first rescues under difficult conditions, and started to evacuate those victims who had not left their homes in time. At 5.30, the crisis committee issued a first report: 600 people had been evacuated before the flood hit, 100 were rescued during the flood, (mostly by boat) and about 300 people were still in their homes. A number of people had to climb to their top floors or roofs to escape the rising water. At around 7.30, helicopters arrived and they began carrying people to safety. Evacuations were also carried out by boat.

At the same time, the question of evacuating 69 patients, including 35 bedridden, from the Bollène hospital was considered. The cellars and ground floor of the hospital were flooded, cutting off the heating and forbidding use of the kitchens. An evacuation plan was developed and at 10.25, the public health and safety department and the hospital director jointly decided to evacuate the residents. The evacuation, supervised by firemen and DSC intervention unit personnel, occurred in the afternoon. Boats and tracked vehicles were used to move the patients who had medical identification tags around their necks. The operation went smoothly and the residents remained calm in the face of this exceptional situation. At about the same time, the body of a Belgian, who had been living in a community for the poverty-stricken, was found in a flooded former convent.

The flood level had barely started to fall on Friday 1 October when the mayor challenged the community to clean up the town in 48 hours. He viewed this activity as a way to avoid depression. He also explained that the worst had been avoided thanks to the actions taken after the Vaison-la-Romaine drama, even though some of the anticipatory measures had proved to be insufficient. In the middle of the afternoon of 1 October, the water level started to drop throughout the area. New bad weather forecasts forced the various authorities to stay on alert, and the mayors expected the worst if new spells of rain fell on the saturated ground, but the threat did not materialize and the situation stabilized.

Crisis management in the Ardennes in 1993[15]

The mayors of the many communes affected and the municipal and departmental fire services were central actors during the rescue operations. The local crisis committee was already in action in the Charleville-Mézières town hall before the committee of the prefecture was established. The fire service set up a temporary departmental fire and rescue operations centre, once it appeared that the Meuse would continue to rise (the Ardennes is one of the few departments where there is no permanent CODIS). The crisis committee in the prefecture was set up on Tuesday 21 December. Because of the rain that had fallen since early December, departmental services had already been placed on alert on 13 December. There had been frequent contact with the mayors of communes subject to flooding. Communication between the municipal and departmental crisis committees was also facilitated by the use of local elected representatives participating in the crisis meetings at the prefecture. The prefecture's building was flooded along with approximately 20 other public buildings. The ORSEC plan was not implemented. It would appear that enough mean has been mobilized on the groudn during the course of the flood without having to resort to the ORSEC plan.

Nearly 2,000 people were moved, but prectically all were taken in by relatives and friends. The accommodations centres only took a limited number of people in temporarily. Considerable solidarity was observed between members of the population and rescue workers alike.

On Sunday 19 and Monday 20 December, the water rose inexorably. Firemen and the staff of the Charleville-Mézières municipal technical services were increasingly called upon to distribute breeze blocks, provide shelter for the most exposed, and pump water. Detours and road closings multiplied. The state public works complained afterwards that it had had to answer about 4,000 phone calls from individuals while they believed the provision of this information to be the responsibility of the mayors. It also complained that the inhabitants, who wished to travel in the stricken town (because of the Christmas festivities) paid little attention to the signposts and sometimes even moved them.

On Wednesday 22, the Meuse reached 5.14m at 8.00 in Charleville and steady rain continued to fall. In one nearby village, 450 houses were flooded and 20 patients were evacuated from a clinic. In another, supplies were ferried by boats requisitioned by the municipality. In yet another, a major bridge and a hydroelectric plant were threatened. Traffic was disrupted in Charleville-Mézières. At the junction between the two parts of the town, walkways were built on breeze blocks. A large number of inhabitants were evacuated by boat. Factories and shops were forced to close, and 400 people had their phones services cut off. The CIRCOSC in Metz was asked to send reinforcements, particularly fire-fighters.

The main bridge in Charleville was closed in the night of 22-23 December, and then ballasted with 150 tons of concrete. The Meuse reached 5.76m at 8.00, and the technicians feared the bridge would be swept away. The Charleville bypass remained the only means of access to the town centre but it was plagued by traffic jams and threatened by flooding. Similarly, rail traffic could no longer reach Charleville. Over 200 people were evacuated during the day, including 180 residents of two elderly citizens' homes. Most of the evacuees were moved to another hospital in the town. Private individuals were helped and provided with supplies by soldiers using rowboats. Approximately sixty people were relocated to a play centre where they were cared for by the municipal service for social action which supplied hot drinks and food before they went on to stay with relatives.

The Meuse reached its peak level in Charleville at 16.00 on 23 December (5.78m). At 18.00, the level appeared to stabilize, and a drop was observed throughout the basin. Accompanied by representatives from the Gendarmerie and the army, the prefect and a deputy visited the stricken area on 23 December to observe the damage in Givet, one of the most seriously affected communes. On 24 December at 14.00, the Meuse stood at 5.70m. The SAC forecast a possible rise in water level on 26 or 27 December. Owing to the slight drop in the water level, crisis operations slowed down a little. A detachment of firemen from the Yonne and a few volunteers from the south of the department remained in service over Christmas, but they had little to do since the inhabitants attempted to have a quiet Christmas despite the flood.

The road between Givet and Charleville was re-opened on 25 December, and on the 26, the water level seemed to be dropping further. Demobilisation of the reinforcement firemen began, and the municipality went into preparation for the post-flood phase. The municipal services began to clear the roads and public buildings. In addition to two-day food parcels (delivered to people's homes at their request by the soldiers), water disinfectant products were issued, particularly in the Mézières district where a supply truck was parked.

Crisis management in the Ardennes in 1995

Like the 1993 flood, the 1995 flood was managed at the communal and departmental level. This time, however, the ORSEC emergency plan was activated several days after the crisis began. The municipal crisis committee, set up as of Monday 23 January, met twice daily until 7 February. The town hall general secretary met twice daily with the prefectoral crisis committee which was set up on 25 January following a deterioration of the situation (the Meuse was at 5.18m at 8.00). A central command post was also established around the town hall general secretary for the management and deployment of various equipment (such as passes and requests for accommodation).

In the prefecture, the interagency crisis committee was set up on the fourth floor of the building (which was fortuitous since the building was ultimately flooded to a depth of 86 cm). The committee worked on a three 8-hour shift principle. Each service within the committee operated as a small separate committee with a relative amount of autonomy. The committee produced reports and passed the information on to both the elected local representatives and to the media. It appears that the media were highly visible and persistent in their requests for information. Despite certain communication efforts such as a bulletin board with the most important data, the staff at the prefecture and on site spent a substantial amount of time answering their questions.

The experience and habits derived from the 1993 flood played a structuring and formative role. Apart from the prefect, who took office shortly afterwards, the main actors had been present during both events. They knew each other and had taken part in various evaluations and reorganization meetings during 1994. During the 1993 flood, the rescue actors, and the fire brigade in particular, had been able to test the condition of their equipment (such as boats) and afterwards had begun to make the necessary adjustments. Likewise, the Charleville-Mèzières municpality developed a disaster plan and so-called 'reflex-cards' for disaster personnel based on the lessons learnt from 1993. These contained practical action guidelines, for example, 'at such and such a level, reinforce this bridge.' The public works department made an inventory of the roads likely to be cut off, and produced road signs for more efficient signposting of the alternative routes. The bypass, which is the last means of access to the town centre in case of flooding, had been improved. The Gendarmerie's role in case of flooding was clarified and EDF had moved some installations away from flood areas, restructured electricity networks, and upgraded its emergency procedures.

Until the morning of Sunday 29 December, the prefect felt he had the necessary means to deal with the situation and was not planning to implement the ORSEC plan, but the situation kept worsening (the Meuse had reached 5.95m at 20.00). The operational actors who had been working at the crisis site for several days under difficult conditions were beginning to feel the strain and needed to be replaced with fresh reinforcements. Consequently, the prefect declared the ORSEC plan to be operational at 22.00 on 29 December. The decision can also partly be explained by the fact that a special inter-ministerial meeting was to be held the following day in Paris to discuss the floods and by the Prime Minister's expected visit the day after that.[16] The use of the plan highlighted the seriousness of the situation, and dramatized the need for central government assistance. On Tuesday 31 January, the prefect visited the area with the Prime Minister, and then again by himself on 5 February, to observe the damage and monitor the progress of the rescue operations.

The scope and size of the operations during the 1995 floods become clear from the following data. From 23 January to 6 February, the firemen of the Ardennes carried out 7,087 interventions, e.g., half of the total number of their interventions in a normal year. From the Ardennes, 2,150 firemen were mobilized on a rotation basis from 23 January to 12 February. They were reinforced by over 125 firemen from other departments and, when the water level began to drop, by a fire and clean-up detachment of 20 men with special cleaning equipment. Local means were supplemented by the municipal services and by reinforcements supplied on request by the Eastern Defence Zone (vehicles, boats, motor-pumps, a helicopter, and a water treatment station).

Unlike 1993 when its role was relatively limited, in 1995 there was considerable military assistance from the beginning of the flood through mid-February. Around 1,800 soldiers constructed walkways, tied them down to withstand the current, and helped ballast bridges. They also ferried a large proportion of the supplies by boat to those who had stayed in their homes, secured fuel and motor oil containers (particularly 40,000 litre containers in companies), gave considerable logistical support to the work of the various public services (EDF and France Telecom in particular), and enabled the population to move about by placing boats or lorries at their disposal to provide shuttle services. Five-hundred gendarmes and 250 police officers were mobilized until late February. They carried out reconnaissance missions along the main roadways, placed no-access road barriers and detour signs and helped drivers find their way.[17] A large number of personnel from the public works department also intervened, as did employees from the many other national services and a range of voluntary associations.

On Monday 23 January, 253 houses in the department were hit and many roads were closed. On 24 January, further heavy rain was announced by Météo-France, and the SAC forecast a worsening of the situation. Rescue operations to protect people and property increased. Walkways and detours were placed at an increasingly fast pace. Over 8 km of walkways supported by breeze blocks were installed during the flood by the Charleville municipal technical services and some 15,000 breeze blocks were used. On Wednesday 25 January, the Charleville municipality decided to open the play centre to provide accommodation and took stock of the available replacement housing.

On 26 January, the town was divided in two by the flood. The public transport services had to re-route their buses and set up a special telephone service to keep the public informed. The mayor called for extreme caution. Municipal teams made rounds in the town to assist the elderly and the handicapped. An elderly citizens' home and the care centre for the aged were evacuated as a precautionary measure.

The slight lull that occurred on Friday 27 January when the Meuse dropped to 5.19m at 16.00 was short-lived as rain continued to fall. More roads were closed and more companies were flooded. In Charleville-Mézières, 300 houses and cellars were flooded, two secondary schools were closed and another elderly citizens' home was evacuated. The administrator of the Corvisart hospital contacted the local crisis committee to place Charleville-Mézières medical teams on stand-by for a possible evacuation of his premises. On Saturday 28 January, the Meuse reached 5.82m at midnight, thus exceeding the highest level recorded in 1993. The fire service stepped up its operations. New walkways were erected, schools were shut and private individuals were evacuated by the army. Fourteen patients from the intensive care/cardiology unit of the Corvisart hospital were evacuated by night, which made the operation more hazardous, and were placed in other hospitals in the region. Though local radio stations, the prefect and the mayor warned the public's about health hazards connected to the flood. Advice was given to avoid certain forms of food, and to follow waste disposal procedures.

On Sunday 29, the prefect issued a decree banning car and pedestrian traffic across three bridges in Charleville, closed certain administration centres and asked the inhabitants to limit their movements. The bypass was closed on the afternoon of 29 January, except to rescue vehicles and public transport. In the evening, the prefect activated the ORSEC plan. In the entire department, one death had been recorded on 27 January in Chooz when a man died of a heart attack as he tried to carry a bag of cement to dry ground. Seven people were injured, 9,932 people were affected by the flood, 634 were evacuated, 3,296 homes were without electricity, 56 had no telephone service, 200 had no drinking water and 4,100 children had no schools to go to.

On Monday 30 January, the town was in dissarray. At 6.00 the Meuse reached 6.06m. The firemen of the Charleville-Mézières had to move out of their station and set up elsewhere. The prefect gave the order to evacuate all of the patients of the Corvisart hospital (123 people) to other hospitals in the region at 12.00. A large number of shops and department stores, the Citroen factory (which supplies 80% of the engine blocks produced by this national car manufacturer) and many others had to close.

On the same day, an economic committee and a "local collective services" committee were established by the prefecture to evaluate the damage done to industrial, commercial and road infrastructures, as well as to the public services. On 30 January, 257 communes out of 463 were flooded, and two were cut off (Anchamps and Ham sur Meuse, where four tons of food requisitioned by the prefect were ferried by helicopter). In these communes, twelve-thousand people were affected, 2,750 of whom were evacuated even though only 536 had to be lodged. Five hundred people from the Givet zone were evacuated to Belgium. Cooperation with the Belgian authorities was good despite the fact that no formal preparations had been made for such a contingency.

On Tuesday 31 January the situation was still critical. The highest water level (6.30m) was recorded after 14.00 and lasted the remainder of that afternoon. Thirty kilometres of road were flooded in Charleville-Mézières, 50% more than in 1993. Firemen provided a boat shuttle service to transport the members of the crisis committee since the prefecture could no longer be reached in any other way. The Prime Minister announced that a state of natural disaster would be declared as quickly as possible, and called for a public body to be created to improve the drainage of the Meuse. Following the Prime Minister's visit, the flood became a political issue. Because there were subsidies at stake, and the Prime Minister himself was a candidate for the May 1995 presidential election.

The Meuse started to drop on Wednesday 1 February. EDF switched the power back on for 3,900 customers which also meant the restoration of heating for some. In Charleville the bypass was re-opened to light vehicular traffic. The director of the municipal technical services was unhappy with this decision because the prefecture banned public transport from using this road. This provides a clue to the tension and differences of opinion that can exist between the prefecture which, within the framework of the ORSEC plan, made all decisions, and the municipal staff who claim to have a better knowledge of the area and of the population's requirements.

The water level continued to drop on Thursday 2 February, but did so at a slow pace (5 cm/hour). In Charleville, the Meuse still stood at 5.78m at 16.00 which was the highest level in 1993. In the department, 1,410 homes were still without electricity, 3,000 homes had no telephone service and 43 roads were still unusable. Traffic was still banned on the three bridges in Charleville, but one lane of the bypass was open in both directions. The army pumped water out of flooded cellars and shops. It also supplied the flood stricken populations with drinking water, especially since the public health department warned against the safety of the drinking water supply via the local press and radio and television stations. Most schools (of which about 100 had been closed) opened their doors again.

The prefect paid his second on-site visit on 5 February and was confronted with 150 demonstrators from a factory which had been flooded and was in danger of being closed. The ORSEC plan was gradually deactivated as of 3 February even though the prefectoral crisis committee remained operational until 10-12 February. Eleven communes (58,000 inhabitants) still had no drinking water and had to be supplied by means of a large drinking water treatment and production station. There were no longer any problems with electricity by 4 February. The prefecture did not open until 6 and 8 February after being cleaned by the army. Municipal workers from Reims, Châlons, Limoges and Nordhausen came to take part in the initial clean-up and technical restoration.

4 Recovery and damage compensation

The natural disaster insurance system

A legal basis — The significant factor regarding compensation in France is the existence of a legal system which combines insurance and solidarity. It essentially provides compensation for property damage (including the property of local collective service organizations whitin certain limits) and company damage. The preamble to the 27 October 1946 Constitution which reappears in the 1958 Constitution, established the principle of 'the solidarity and equality of all French citizens facing the expenses incurred through national calamities.' This principle remained nothing but an assertion for a long time with only limited manifestations in terms of damage compensation. It was not until the law of 13 July 1982 relating to the compensation of victims of natural disasters that obligatory guarantee measures were implemented. This law, the cornerstone of the system, was complemented by the law of 25 June 1990 which modified the insurance code and extended the compensation scheme for compensation of natural disasters to the Overseas Departments, and the law of 16 July 1992 intended to adapt the legislation relevant to insurance and credit to the European single market. These legislative measures are property compensation measures and do not cover individuals.

There is no strict and precise definition of the notion of a legally insured natural disaster nor is there a list of events excluded from the insurance. The disaster is evaluated using different criteria such as the "abnormal intensity" criteria (defined in the 25 July 1990 law) and the criteria of "uninsurability of the risk" which is stipulated in the 16 July 1992 law and which states that only 'uninsurable direct material damage, the determining cause of which is the abnormal intensity of a natural agent' is covered under the natural catastrophe scheme.[18]

The scheme instituted by the 1982 law was created under the pressure of events, particularly the floods in late 1981 – early 1982 which affected the Rhône and Saône valleys and the South West of France, and was inspired in part by the system established in 1964 for agricultural disasters. In some respects, the scheme resembles an insurance scheme since it is founded on the principle of individual willingness to be insured and is established according to current insurance premiums (damage to property, cars) which make mutual benefit risk insurance possible. In other respects, it resembles solidarity since it is organized by the state. In the first place, the implementation of the scheme depends on the public authorities whose task it is to issue an inter-ministerial order declaring the 'state of natural disaster' according to a strict procedure. Second, the premiums are set as a percentage of reference premiums and not, as is usually the case, according to the potential value of the damage. And third, the state gives its guarantee to the *Caisse Centrale de Réassurance* (CCR, central fund for insurance) for the risks caused by natural disaster.[19]

An elaborate procedure — To implement the scheme, there must be a state declaration of natural disaster via an inter-ministerial order. This declaration is made by means of a process involving different steps and levels. A request is formulated by the mayors and sent to the departmental prefect who has a month to complete a departmental file including a report on the nature and intensity of the catastrophe, a technical report compiled by the competent services (particularly *Météo-France* in cases of flooding), a list of the affected communes with, if possible, a map to locate them and all documents showing the abnormal intensity of an event (press cuttings, photos). The file can deal with one or several communes.

This file is then sent to the DSC at the Ministry of the Interior and is submitted to a committee for advice. An informal Commission meets to examine each event. It brings together representatives of the DSC, the Ministry of Economy and Finance, the Ministry of the Budget and the Ministry of the Environment, and its secretarial support is handled by the CCR. The committee produces a recommendation which may be transformed into an official disaster declaration. In this occurs, the declaration is made the subject of an inter-ministerial order which is signed by the Ministry of the Interior, the Ministry of Economy and Finance and the Ministry of the Budget and is published in the *Journal Officiel* (government bulletin).

Subsequent to the July 1992 law, the inter-ministerial order must also determine the zones and periods of the disaster as well as the nature of the damage caused. The committee sometimes has difficulty reaching a decision. The file may not be complete, the very notion of 'disaster' is not always clear or easy to specify, and above all, the extent of the damage observed is not a criterion in itself. The process can be speeded up by ministerial requests. If the decision is negative, the DSC – though not obliged to do so – sends a telegram to the prefect informing him of the decision and giving reasons for it. The prefect may sometimes modify and complete the file. Only the supreme administrative court can judge whether the state was wrong not to recognize the condition of natural disaster.

The publication in the *Journal Officiel* influences some of the obligations of the policy holder and the insurer. The policy holder is obliged to declare any damage likely to be covered by the insurance as soon as he becomes aware of it, and not more than 10 days after the publication of the inter-ministerial order for direct material damage (not more than 30 days for running losses). The insurer is obliged to pay compensation within three months after the insured party submits the estimate of the damaged property, or if the publication of the order in the *Journal Officiel* is later that the damage estimate, within three months of its publication. It should be noted that in cases of disaster, insurers set up rapid reimbursement schemes and that this insurance system does not exclude other specific actions which the state may decide to adopt.

The prefect has a key role in the compensation process. A circular of 26 March 1993 stresses that 'the prefect coordinates the post-crisis follow-up. After the emergency operations, the aid given to the population changes in nature but remains indispensable for meeting material needs (e.g. rehousing and restoration of surroundings), providing medical assistance, getting business started again, ensuring payment of those who provided services in the emergency phase, dealing with litigations of all sorts and continuing to inform and orient the victims.'[20] Case by case systems are set up between state and municipal services, and the charitable associations for the sharing of emergency aid among the victims.

Comprehensive coverage? — The scheme only applies to those who are insured. It is not a specific type of insurance underwritten intentionally for this kind of damage. It is connected to the basic insurance policy, usually fire insurance, and it covers the costs of clean-up, demolition, pumping and cleaning and, since the 1992 law, the costs of geotechnical studies for damaged buildings. If a running losses policy has been taken out, the insurance covers the loss of profits before taxes and the extra running costs during the compensation period of the policy. The important condition to qualify for compensation is that the damage be the direct consequence of the disaster.

There is no relation between the premiums fixed by the public authorities and the degree of exposure to risk. The risks are spread and an element of solidarity is applied across the nation. The scope of the underwritten insurance in the basic policies (property, homes, automobiles) determines the amount of the compensation. This is why some compensations were considered insufficient in relation to the damage sustained.

This system has worked relatively well until now from both the insurers and the insured parties' points of view. The insurers have always made a profit since the amount earned in premiums has always been higher than the amount paid out in compensation. Another indication that the system works well is that the insurers are making less use of the central reinsurance fund guarantee. Thus, Serge Magnan of the *Fédération Française des Sociétés d'Assurance* (French federation of insurance companies) was able to state that from 1982 to 1993, 'receipts have reached FF 39.2 billion in premiums for 20.3 billion in payments,' while adding that although the system is viable it does not have large reserves.[21] In fact, some anxiety exists due to the frequency and seriousness of flooding in recent years, particularly since 1993. There are a few cases of policy cancellation by the insurers in places where the risk has already occurred and may occur repeatedly (one aim of the insurers is to have preventative measures taken before accepting to insure again).

The situation for the policy holders also appears quite good. Conflict occurs in less than 1% of the cases. More often than not, disagreements are about the amount of compensation, especially for shopkeepers and craftsmen who underinsured their stock or equipment. Between 80 and 90 percent of the files are dealt with within the time legal limit (3 months), and only cases of company compensation can take longer to resolve since they are more complex, particularly in terms of running losses.[22] Companies roughly account for 50% of the compensations in terms of value, and the rest is divided between private individuals and communes.[23] It is believed that only 30% of companies are insured against running losses and that approximately 5% of private individuals are not insured (on average and with regional fluctuations). Finally, houses are insured for rebuilding costs only if this is done within two years of the disaster and on the same site as the former home – which is problematic in a zone known to be a flood risk area.

The 22 July 1987 law (article 13) limits the reimbursement of those expenses directly incurred through the emergency operations. In theory, this law obliges the beneficiary commune to reimburse the cost of rescue means deployed by the state, and municipal services and other public bodies. However if the ORSEC plan was declared, the beneficiary commune repays nothing as long as the other public agencies which provided aid belong to the same Defence Zone or to a region threatened by the same type of risk as the one in question. In fact, the repayment principle has never really been put into practice, and the exceptions established in the law have been reinforced by practice. Thus, it seems that the state or even the operating services, out of solidarity and a desire not to aggravate local situations, exempt the beneficiary communes from this repayment. Also, exceptional subsidies can be awarded by the state to alleviate the financial strain.

The legal compensation system does not exclude other forms of aid and subsidies which may come from municipalities, departments, regions, the state or the European Union. Even if compensation can be paid rapidly, emergency aid for private individuals is often required, particularly for re-housing, clean-up and renovation. Also, the compensation provided for by law only covers a part of the damage caused to public collective services and their property. In fact, local collective services do not insure all their infrastructures because of the inappropriate damage insurance system provided. For example, there is a problem of insuring against fire damage to large infrastructures such as bridges and roads.

Damage compensation in the Vaucluse in 1993[24]

The compensation phase started in Valréas and in Bollène on the afternoon of 1 October. A layer of muddy, silty water had ravaged the two towns. Priority was given to the clean-up and the sorting of damaged property. In both towns, the municipal and departmental fire services, the soldiers of the intervention units and the municipal technical services devoted their efforts to pumping, clearing and washing, especially on roads and in buildings. Personnel from EDF, France Telecom, the civil engineering and water departments, and requisitioned public works companies worked around the clock to put the major collective services back in operation.

Private homes and shops were tended to with the help of 1,000 to 2,000 volunteers who cleared away mud and hosed down or threw out furniture and personal belongings. Inhabitants from neighbouring communes came by to help. This was much appreciated by the locals, who were in a state of shock because they had lost most of their property and because of the extent of the enormity of the clean-up work facing them. Inhabitants of Vaison felt that they had overcome their own disaster in 1992 thanks to similar solidarity and came to give their help. In Bollène, a special form of organization was set up. The gendarmes closed the town to prevent looting and facilitate the cleanup. Inhabitants and volunteers had to go to the town hall to obtain a pass consisting of an ink-stamp on the arm in order to enter the town. The volunteers were then organized in teams according to each district, and each team was led by a municipal employee or a local fireman. In both towns, the municipality and voluntary associations distributed cleaning equipment.

Among the rare sources of information concerning the damage, one can refer to the estimates compiled by the municipalities after the crisis.[25] In Valréas, the victims included 460 individuals (for an estimated total of damage of FF 23 million), 35 shopkeepers, craftsmen and industrialists (FF 92 million) and 50 farmers (FF 3 million). The damage incurred by the commune (roads, networks), was to be estimated at the end of 1993 was FF 89 million. This results in an estimated total of FF 207 million. Nearly 400 homes were damaged, a hundred of which could not be occupied immediately, and 38 to 40 families were completely homeless. All of the families who filed an application with the municipality were re-housed within 24 hours.

In Bollène, there were 2,410 flood victims. Around 131 families had to be re-housed, and nearly 200 businesses were more or less damaged. These businesses included 40 farmers, 90 craftsmen and shopkeepers, 33 other diverse businesses and 30 self-employed people. In Bollène, the total amount of emergency aid awarded to individuals amounted to FF 6,250,000. The damage incurred by the commune was estimated at FF 45,747,000. In both towns, several tens of kilometres of river banks, departmental and national roads, by-roads and about 10 bridges were seriously damaged and in need of repair. Similarly, large areas of farmland on the territory were deeply furrowed and fields and vineyards were covered in a layer of mud – which makes farming difficult for a long time. Ditches and access ways were destroyed.

The compensation procedure established in the 1982 law helped to structure the compensation process. Reception centres were set up in both town halls, staffed by about thirty volunteers in the case of Valréas, and by municipal administrative personnel with free help from insurance experts, a chartered accountant, a financial advisor, an agricultural advisor and a lawyer in Bollène. The victims submitted a disaster claim which was added to the file that each municipality swiftly passed on to the prefect with a view to declaring a state of natural disaster. This formal step served to make people aware of the extent of the losses and, at the same time, provided a more concrete prospect of compensation for the victims. The procedure was often difficult to manage both for staff and victims given the requests for bills, inventories, and other evidence. The mayors provided extensive information urging people to make claims of this sort.

In Valréas, the mayor warned people against the temptation to have their repair work done quickly via "moonlighting", since this would prevent them from having invoices reimbursed. They were also cautioned against false experts or representatives from cleaning companies wrongfully charging for services. Moreover, the insurance companies soon opened local temporary reception centres and appraisers started making visits to make estimates, and some companies paid out money before the state of disaster was even declared. Appeals to have a state of agricultural disaster declared were made in order to complete the compensation of farmers. The state of natural disaster was officially declared on 11 October for both communes and this marked the start of the 10-day period in which to transmit official claims to the insurers.

More or less standardized emergency aid was also awarded via other formal procedures, and some of these have already been outlined. For example, the Bollène municipality provided funds for the association it set up (*Bollène Solidarité*) while sharing some municipal funds among the victims in the first week (FF 3,000 in cash per household, plus FF 3,000 in cases of a loss of livelihood). A broad public-private municipal coordination commission was also set up in order to distribute the total financial aid allocated for the victims.

In Valréas, the municipality awarded emergency aid to those worst affected via an emergency commission (associating the same kind of representatives as in Bollène), and increased its efforts on all of the supplementary social benefits such as food vouchers, assistance in paying school lunches and heating bills, medical aid and the distribution of clothing, furniture and electrical goods with the help of the private relief organizations. This material donation management also occurred in Bollène with the help of various charitable associations.

Other organizations also provided emergency help. For example, after 6 October, on behalf of the European Community, the Red Cross distributed aid totalling FF 2,000 in cash for each household which had been damaged or to families who had had a member killed or hospitalized, as well as FF 2,000 supplemented by FF 500 per dependant on behalf of the *Conseil Régional*. Also, charitable associations distributed limited financial and material aid. Moreover, the Red-Cross made a significant contribution in distributing drinking water since this was not supplied in Valréas until 6 October. Finally, the government, via the Ministry of the Interior on 2 October and the Ministry of the Budget on 6 October, announced 2.6 million in emergency aid for the victims in the Vaucluse and 120 million for the local collective services in the area as a show of national solidarity.

Another important aspect of the compensation phase concerns the various actions carried out in the fortnight following the floods by a number of institutions to facilitate the living conditions of individuals, businesses, shops and farms affected by the event. The two mayors asked the prefect for exemptions from fees for renewing identity cards, passports and driving licenses which had been destroyed, as well as for a moratorium on the EDF/GDF, France Telecom, water board and tax bills and on bank payments for all of the victims. Similarly, the national health insurance fund was asked to provide full coverage of the victims' health expenses. The Ministry of the Budget was also approached in order to obtain a moratorium on the payment of shopkeepers', craftsmen's and self-employed taxes. The *Conseil Régional*, supported by the communal and departmental elected representatives, also requested that the various partners involved renew and extend a housing renovation operation set up following the 1992 floods in Vaison-la-Romaine in both communes. The guild chamber, after identifying a hundred or so craftsmen who were victims of the flood in Bollène and Valréas, sent two agents from its economic service to meet and advise the craftsmen on how to compile their dossiers in order to obtain financial aid, rebates and payment delays. It also set up a scheme in relation with the various social security funds and administrations in order to obtain measures in favour of the victims, and opened a bank account for 'Craftsmen flood victims' to receive donations. Other professional associations took comparable initiatives.

Damage compensation in the Ardennes in 1993

The retreat of the river was confirmed on Monday 27 December (5.33m at 8.00, 5.28m at 16.00). Municipal services, firemen and soldiers actively took part in the cleanup operations. There were many acts of solidarity among the inhabitants. Sometimes these took the form of a formal, pre-established framework as in Warcq, a frequently flooded commune where a 100-member mutual assistance committee has existed since 1983. Various departmental and municipal authorities distributed disinfectant and chlorine tablets and gave advice on how to disinfect premises, treat the water and dispose of rubbish.

There were many public information initiatives, particularly on the subject of compensation. The Charleville municipality circulated a letter on 27 December to all the inhabitants, reminding them of the necessary formalities. The chamber of commerce appointed one of its members with legal skills as 'flood czar' to inform the flood stricken businesses of the procedures to follow to obtain payment delays.[26] It also produced a document detailing the constraints to be respected in such cases. Press releases were published beginning on 23 December by the general insurance agents' chamber of the Ardennes.

Over 3,000 people were stricken by the flood. One-thousand homes were damaged, 91 businesses had to introduce part-time lay-offs while another 25 waited for appropriate aid and a large number of roads and at least 20 bridges were more or less severely damaged.

The compensation procedure via the declaration of the state of natural disaster was quickly set in motion, and the inter-ministerial order was signed on

7 January 1994. Exact figures on the the total damage are difficult to obtain. Estimates run up to FF 750 million; FF 500 million in compensation is said to have been awarded by insurance companies to private individuals and businesses. Some problems arose in relation to insufficient insurance of private and professional property, insurance needs of low income families, and non-compensation for some businesses that were not directly flooded but had suffered serious revenue losses. As in the Vaucluse, a certain amount of material and aid in-kind was proposed by the municipal social services and charitable associations (food, clothing, furniture, psychological counselling). Funds for helping the victims were also allocated via the Ministry of Social Affairs (FF 100,000), the *Conseil Régional* (Champagne-Ardennes Région) (FF 250,000) and the neighbouring town of Reims (FF 50,000). The distribution of this aid was managed by a committee set up at the prefecture (with representatives of social care services and charitable associations).

The department of Ardennes was awarded special aid by the state to repair communal and departmental infrastructures. Following the various floods in France in December 1993-January 1994, the Ministry of the Interior repeated the procedure it had set up after the Summer-Autumn 1993 floods. On 17 January 1994, the same Ministry of the Interior official was assigned the task of evaluating the damage of the Winter floods and propose implementation plans for government aid measures.[27]

The 1993 flood reinforced political debates about infrastructural works to be carried out along the Meuse. The commune of Warcq took the initiative in 1992 to form a consortium to carry out work designed to bring flood levels down by 40cm. The cost of these works (re-gauging of the river, improving the banks, locks, movable barriers) totalled FF 16 million. In December 1993 the consortium had just received a subsidy of FF 9.9 million from the European Fund for Regional Development to start the work. Questions had been raised about this project and these were reinforced by the floods. Mayors of upstream communes feared that these works would increase the flood level in their own communes. Similarly, a citizens' action group set up in February 1994 criticized the project, claiming it would be more useful to combat certain types of deforestation and forest farming.

At the national level, an Ardennes deputy launched a media appeal for economic support to the stricken area. He called for the creation of a parliamentary inquiry commission to examine the causes and remedies of flooding – a commission that he was later to chair. Locally there was a certain disappointment at the absence of visits from senior government officials. While at the same time the President had visited the flooded streets of Saintes (Charentes) by boat and the Prime Minister had gone to the Camargue when it was hit by a flood, the department of Ardennes was bypassed by the political elites.

Damage compensation in the Ardennes in 1995

The population was faced with extremely difficult circumstances. Many of the inhabitants, individuals and professionals, lost everything for the second year running. Some of them had only just finished repairing their premises. The new damage was extensive and walls of buildings remained damp for many months. Despondency and stress were commonplace. Local flood assistance points were set up on 4 February in accordance with the assistance plan developed by the municipality following the 1991 and 1993 floods. They should inform the population about social assistance, insurance questions and the consequences of the flood, particularly regarding hygiene. They distributed disinfectant, rat poison and served as headquarters for the clean-up teams organized by the charitable associations.[28] They welcomed around 1,850 people in 9 days. As of 26-27 January, the Chamber of Commerce again designated a "flood czar", and the guild association set up a service to assist craftsmen with the administrative procedures in banks, administrations and social organizations. As of 5 February, a natural disaster coordination committee was established by various insurance agencies.

With the help of associations, inhabitants and shopkeepers filled containers with spoiled products and furniture while the army dealt with cleaning up public areas. Mutual help among people, which had already been substantial in terms of housing, continued in a positive manner. The media were instrumental in this process. A 24-hour switchboard system at Radio-France was established, and messages were broadcast and appeared in the regional newspaper. A special radio programme was broadcasted, on 3 February. Individuals and charitable associations were able to propose aid (accommodation, clothing, help with cleaning, donations in kind and in cash) and demonstrate their support. Also, in the context of this operation, as in other more traditional charitable operations, bank accounts were opened to receive donations.

According to the reports issued by the prefecture, the flood produced over 12,000 victims.[29] According to the estimates, 2,000 to 5,000 homes, 1,100 firms and over 10,000 employees (20% of the local economy) were affected by the floods. For companies which had been hit by the flood in 1993, the damage was two to three times greater in 1995 due to the higher water levels. Local services estimated their damage at FF 140 million. The *Conseil Général* estimated the damage done to departmental roadways at FF 46.71 million and to structures worth FF 11 million. Damage to national roads was estimated at over 4 million and 20-25 million to communal roads. Damage to locks and fixtures along the Meuse amounted to 20 million.

Inter-ministerial orders declaring the state of natural disaster for 236 communes in the department of Ardennes were signed on 6 and 21 February

(*Journal Official* of 8 and 24 February). On 25 January, following the floods that affected a large number of departments, a 24-hour service had been set up at the Ministry of the Interior, and the latter had decided on an exceptional procedure to speed up the handling of files.

Unlike 1993, the insurance valuers responded quickly and came from different parts of France. The Managing Director of AGF, an insurance company which is well represented in the area, came to Charleville-Mézières on 8 February to observe the extent of the damage. He declared that his company would mobilize experts so that claims could be paid as quickly as possible, and that everything would be done to provide advance payments. The compensation would reach FF 750 million in 1995.

Apart from compensation, other kinds of aids were granted to firms which were badly affected, especially because of the repetition of the phenomenon one year after the first flood and its damages. Different professional services, the guild chamber, the the Conseil Général set up finds or systems to give exceptionnal funds advances or loans (especially for equipment and material investments), helped to obtain delays for the payment of taxes and social contributions.

Around 560 private individuals were given emergency aid, worth a total of approximately FF 1.2 million francs. From February 3 on, a similar system to the 1993 one was set up for the distribution of this aid. Here again, the municipalities set up special aid systems (for example, the Charleville-Mézières municipality allocated aid covering 12% of the cost of house interior restoration for the un-insured). The direction of the fiscal services for the Ardennes adopted measures in favour of tax-payers (payment dead-line postponements, no penalties). A large number of declarations and public actions can be noted on the part of local elective representatives requesting that the government take the difficult conditions in the department into consideration.

Another mission to evaluate damage was assigned to the inspection service of the Ministry of the Interior. On the basis of the enquiry, an exceptional scheme was adopted by the central government on March 23. The department of Ardennes was granted a subsidy of FF 22.5 million for the repair of uninsured local services infrastructure. The State also awarded aid worth FF 4.8 million for road infrastructures, and FF 55 million for structures necessary for navigation on the Meuse and the Aisne rivers.

Moreover, though the coverage of the flood in the media was appreciated, a communication campaign promoting the image of the Ardennes was launched in the weekly and daily national press shortly after the disaster. This campaign, whose costs were shared by a coalition of four public and private bodies was initiated at business owners' request in order to counteract the negative effect on the public image of the Ardennes caused by the floods.

5 Concluding observations

Assessing and improving risk communication and warning

For flash floods, anticipation appears to rely essentially on weather forecasting. Increasing the number of measuring stations along rivers prone to flash floods would hardly increase the amount of time available to react. In the Vaucluse, where no BRAM or ALARME was sent out by Météo-France in 1993, the post-disaster debate focused on the installation of a new radar facility. Another discussion regarded Météo-France's messages. Severe criticism was levelled at the complete absence of bulletins, the communication of BRAM or ALARME bulletins that turned out to be faulty and the transmission of BMS alerts not followed up on. All this was said to reduce the credibility of warnings and reduce local sensitivity to them.

For riverine floods anticipation hinges largely on the SACs backed up by meteorological data. The flood experiences revealed that number of SACs and staff levels were insufficient. In some regions and departments, flood announcement has been a low priority area, and there was no legal framework outlining their administrative responsibilities. The SACs were often poorly equipped, and the models used to forecast water levels were inadequate. Calls were made for improved cooperation between Météo-France and the various SACs. Agreements and protocols to this effect started being drawn up in 1994.

In the cases studied here, the high vigilance and informal warning initiatives of local authorities went a long way towards compensating the definciencies of the official warning system. Prior flood experiences had enhanced local sensitization. A major problem for local authorities was the speed at which information was communicated. Communications problems occurred between the prefectures and the mayors, with the latter either being alerted relatively late or in cumbersome ways. Since 1993-94, experiments have been conducted using automated systems and voice servers. Mayors however, tend to prefer direct interaction and human contact when it comes to the exchange of such vital information. The experiences of the mayors also led them to demand more precise information, as well as a more readable and easier-to-use presentation formats.

Both in the Vaucluse and in the Ardennes, and despite the significant involvement of the operational services and local elected representatives, a sizeable proportion of the population did not give enough credence to the information and advice. They decided mostly for themselves what the risks were. Many locals acted as disaster tourists rather than as potential victims during the onset of disaster. Taken together, these factors complicated the rescue and evacuation process, raising the prospect of forcible evictions. Tkey issue is how to a productive "risk culture" can be nurtured in communities where floods are a recurrent and/or life-threatening phenomenon.

Assessing and improving flood response

The three cases show that major disasters trigger a complex intergovernmental process. Local initiatives are crucial in the first stages. Next, prefectoral interventions occur, either to facilitate or to partly replace local crisis management strctures. In addition the CIRCOSC, the DSC and the army become involved in coordination and support roles. The ORSEC plan is no longer the only means for mobilizing reinforcements from outside a department or from the army, or for organizing the management of the crisis. A lot of assistance can be arranged informally. The plan's formal implementation in the Ardennes in 1995 did facilitate the rotation of reinforcements during the protracted operations. It also strengthened the role of the prefectorial echelon, but did not serve to centralize the entire process.

With or without the ORSEC plan, initiatives specific to each organization (according to their area of competence, their structure and preparation beforehand) coexist along with a minimum of dialogue and coordination. Planned responses coincide with improvization in the face of extraordinary problems on the ground. Given this inter-organizational character of crisis situations the various actors can no longer overlook the need to cooperate and find complementarities between the means and skills each possesses. The actors involved do not operate on an equal footing however, nor do they all have identical competencies. The prefect, given his prerogatives, remains a first among equals, but his defacto power also depends on the personalities of actors and the situation-specific social and political circumstances in the stricken area.

A number of concrete problems appeared regularly. Some were mainly logistical: a lack of equipment or equipment unsuitable for the risk involved; the poor lay-out of operations rooms (no sufficient division between action space and decision making space, no adequate modern telecommunication means); poor coordination and delays due to a lack of standardization of the various actors' badges; the sometimes insufficient hierarchical rank of representatives from different institutions on the crisis committees; and the absence or insufficiency of officials on duty in various local level services. Other problems derived from the upward flow of information and communication. There were too many correspondents on the ground and on the crisis committees giving information or formulating requests to superiors, and there was a lack of clarity in the transmission of operational information by the actors on the ground. Moreover, there was a rather strong tendency among various organizations and services to 'go solo' and operate quasi-autonomously, even when they were part of joint crisis coordination committees. Finally, media relations were often ill-structured.

Recovery and damage compensation

A striking feature of the post-crisis phase was the widespread solidarity displayed by all actors involved and the fast and broad based mobilization of resources on the ground. The smooth process of making large networks operational again (water, electricity) and the cleaning-up of public and private premises bear witness to this. Funds were released by a large number of municipal, departmental and regional administrations and various institutions. A number of actions at national and local levels were taken to overcome payment delays and obtain facilities for the flood victims. A major gap in the recovery process was the lack of provisions for coping with the indirect and immaterial damage, such as psychological problems and post-traumatic stress, immediate and long-term health problems linked to spending time in water and living in homes that remain damp for long periods.

The insurance compensation procedure turned out to be a vital if not wholly undisputed element of the recovery process. Insurers have been suspected of profiting too much from the system, of not always acting with due speed, and of imposing restrictions that verge on the illegal – such as the temporary or definitive cancelling of policies in the risky zones. On the whole the system works well however, and in the wake of the floods the insurers did make a major effort to hurry experts to the area, to inform policy holders and to pay advances.

The question of damage to public property has not yet been properly solved. The aid given by the state seems insufficient to enable a complete restauration of public infrastructures to pre-disaster levels. The municipalities, *Conseils Généraux* or even *Conseils Régionaux* are forced to make major contributions to recovery over a number of years.

In the wake of the floods there has been a renewed emphasis on flood prevention in France. Many national and local initiatives have been undertaken. Mayors and prefects again assume a prominent role here. At stake are the conditions for granting building permits as well as the elaboration and updating of the plans for land use, of risk cartography and of prevention plans. The Ministry of environment in turn is focusing upon the redevelopment of rivers and river beds. A project to create a public body for the development of the Meuse and its tributaries (EPAMA) is underway, involving a number of departments across two Regions. In 1995, the local elected representatives in Warcq and Charleville-Mézieres invited their European counterparts, particularly the Belgians, to set up an European Association for the Development of the Meuse in order to deal with the causes of flooding and upgrade their joint prevention and mitigation capabilities.

Notes

1 This study was carried out with the collaboration of the *Direction de la Sécurité Civile* (Ministry of the Interior), the prefectures and municipalities concerned, as well as the various operational services and organisations involved. It is the result of enquiries based on the analysis of various documents (non-official, informal notes and reports, press reports, audio-visual documents, internal memos, laws, decrees and various orders, official reports, etc.) at both the national and local levels, and on interviews in the two departments concerned.
2 The regional prefect has no hierarchical power over the other prefects but plays a coordinating role.
3 The deconcentrated state services are the local agencies of national ministries.
4 Inter-ministerial order of 27 February 1984 on the re-organisation of flood warning and the communication of flood warnings, and inter-ministerial order of 27 February 1984 on the re-organisation of the flood-warning services. *Circulaire* of 2 September 1993 relating to communication of weather alerts.
5 We have translated *établissement public* in 'public body' which refers to a specific public status that can be used to manage some public utilities (it is different from the status of an administration).
6 The SACs must communicate their data to the meteorological services so that these can improve the accuracy of their forecasts. Conversely, the meteorological services assist in the flood announcement by sending them information. When one or more thresholds are exceeded, the meteorological services must provide the head of the SAC and the prefect with weather bulletins at fixed intervals.
7 Concepts placed in quotations in this section refer to administratively designed states of mobilization.
8 Article L. 131-2, paragraph 6 of the *Code des Communes*.
9 Note that the SDIS are under the authority of the prefect from an operational point of view, but from a budgetary point of view in particular, they are answerable to the *Conseil Général*, an assembly elected at departmental level whose powers have increased considerably following the decentralisations started in 1982.
10 Ministère de l'Intérieur – Inspection Générale de l'Administration, *Le préfet et la crise*, *Charte du comportement*, juillet 1994.
11 Note that there can be several warning, reception and processing centres in some departments shared between the various fire services. Also note the existence of another rescue phone number (nr. 15), run by the *Services d'Aide Médicale Urgente* (SAMU, emergency medical aid service) which belong to hospital centres.
12 Article 22 of the decree of 6 May 1988.
13 It is worth noting that not all the zones are equally vital, neither do they all have truly operational and experienced structures. Owing to the large number of forest fires, the southern departments were the first to have a CIRCOSC in 1979 eight years prior to the law. This unequal structuring may play an important role, especially where cooperation between operational participants is concerned, according to the geographical localization of the disasters. The centers have around 90 people, firemen or soldiers, at their disposal.

14 Vincent Dye, *La sécurité civile en France*, Paris: PUF 1995.

15 It is difficult to obtain precise information which can be cross-referenced concerning the management of the rescue operations in the Ardennes in 1993. Apparently, few documents have been produced by the various services concerned. The people interviewed, who took part in both flood operations, have forgotten the 1993 flood, confuse it with the flood of 1995 or make no distinction between observations that can apply to both cases, focusing automatically on the latter event.

16 All of the ministers concerned by the bad weather and floods that continued for over a week convened on 30 January to evaluate the situation, the implemented means and the necessary aids.

17 As in 1993, these actors, and especially the DDE, complained about the lack of cooperation demonstrated by individuals who ignored signs, moved them and resented the security forces for forbidding access to bridges. These reactions forced the actors concerned to mobilize more personnel than necessary, especially to handle the curiosity seekers who wanted to observe the flood often found themselves in danger spots. Gendarmes and police officers conducted rounds and closed some zones to prevent pillaging and break-ins.

18 Yvonne Lambert-Faivre, *'L'indemnisation des victimes de catastrophes naturelles,'* Communication au Colloque 'La crise et le droit de la sécurité civile,' Avignon, 14-15 mai 1993.

19 CCR: a limited company, 100% state-owned since 1 January 1993. An agreement with the state allows it to insure natural disasters, with an unlimited guarantee. The insurers can choose other insurance companies as well.

20 *Circulaire* nr. 93-00104C relating to risk and crisis management: organization of inter-ministerial defence and civil protection services, Ministry of the Interior and Public Safety, 26 March 1993.

21 Vittoria de Bagnolo, 'Inondations: prévenir plutôt que guérir', *L'Assurance Française*, nr. 705, janvier 1995.

22 Ibidem.

23 A special set of measures exists for farms, established in the law of 10 July 1964, which covers agricultural disaster – including floods.

24 It is difficult to obtain accurate and appropriate numerical data. No specific organization is in charge of collecting data and practically no organization or service compares its data with that of other institutions. The presentations made by the various institutions bear no common format in terms of categories of damage, categories of subsidy, etc. Even for the same type of institution in different locations, the number of documents available can vary tremendously. Also, the data tends to be aggregated with other data which either corresponds to a set of floods over a given period or to different floods affecting the same zone. This is just as true immediately after the event as it is when reports are written later. In addition, immediately after the event announcements are made concerning the allocation of a certain number of subsidies, but later it often becomes difficult to find clear signs of the translation of these announcements into concrete action. Some of the announcements can be equated with a public display, while others find themselves integrated in less specific financial procedures than those covering only flood damage (for example, integration of the repair costs caused by flooding within the overall annual budget for the maintenance of local collective infrastructure). The question of reproducing an exhaustive assessment is raised as well as that of an homogeneous comparison of several floods, and hence the differences in content for the three cases.

25 Cf. *Le Tour de Ville. Revue d'information de l'Association Valréas Communication*, nr. 8, 4ème trimestre 1994 et nr. 10, 4ème trimestre 1994. *Bollène. L'information municipale en direct*, 'Spécial inondations. 1er octobre 94 : un an après', édité par la ville de Bollène,

supplément au nr. 7. *Bollène. L'information municipale en direct,* 'Spécial: 1993...94-95, le rebond', septembre-octobre 1995

26 Professional consular organization representing industrialists and shopkeepers at the departmental or local level and under ministerial supervision.

27 *Rapport sur l'évaluation des dégâts causés en France par les inondations de l'hiver 1993-1994*, présenté par Mrs Gilles Sanson et Tristan Florenne, Ministère de l'Intérieur et de l'Aménagement du Territoire, juillet 1994.

28 For the department, this includes especially the Red Cross, the *Secours Populaire, the Secours Catholique* and the *Restos du coeur.*

29 Cf. Préfecture des Ardennes, Direction des Actions Interministérielles, *Les inondations*, 22 février 1995; Préfecture des Ardennes, Direction des Actions Interministérielles, *Dispositif d'aide aux particuliers, entreprises et collectivités locales victimes des inondations de janvier 1995*; Préfecture des Ardennes, Direction des Actions Interministérielles, *Conséquences économiques des inondations*, septembre 1995.

Chapter 3

FLOOD MANAGEMENT IN BELGIUM

Hugo van Hassel and Leen van Lindt

1 Introduction

Two times in recent years, Belgium has been severely affected by rising river levels and flooding. The 1993 floods began on 20 December when, due to continuing rainfall, some rivers could no longer accommodate the increasing volumes of water. The situation returned to normal gradually during the first few weeks of January 1994. Almost exactly a year later, many areas were again affected by rising river levels. This time the situation escalated on 26 January 1995 following a period of intense precipitation. On both occasions, the flooding led to the launching of an impressive, large-scale operation. Over a period of about two weeks, thousands of volunteers, emergency personnel and all available resources and equipment were deployed to relieve the worst effects of the disaster. In addition to causing a great deal of misery and inconvenience to local inhabitants, the floods were also responsible for several billion franks' worth of damage. The River Meuse (in Flanders referred to as the river Maas) and its tributaries, in particular, caused severe problems in the provinces of Limburg, Namur and Liège.[1] The provinces of Eastern Flanders and Luxembourg, and to a lesser extent the provinces of Western Flanders and Hainault, also experienced their share of problems with flooding.

The 1993 and 1995 floods affected different, geographically distinct areas spread across no less than seven provinces. Because disaster management operations were organized at the provincial level, the areas of the study are divided accordingly. In order to provide a more in-depth study of disaster management, we have focused on the provinces of Limburg and Namur in our analysis of the actual course of emergency operations. Both provinces were badly affected by the flooding both in 1993 and 1995. They are part of two distinct regions, the Wallonian and the Flemish, that both have their own administrative structures, contingency plans and disaster management policies. In addition, they both share borders with the Netherlands and France, which are also involved in this International Flood Project. However, wherever possible, we have tried to discuss the progress of the crisis management at a general level.[2]

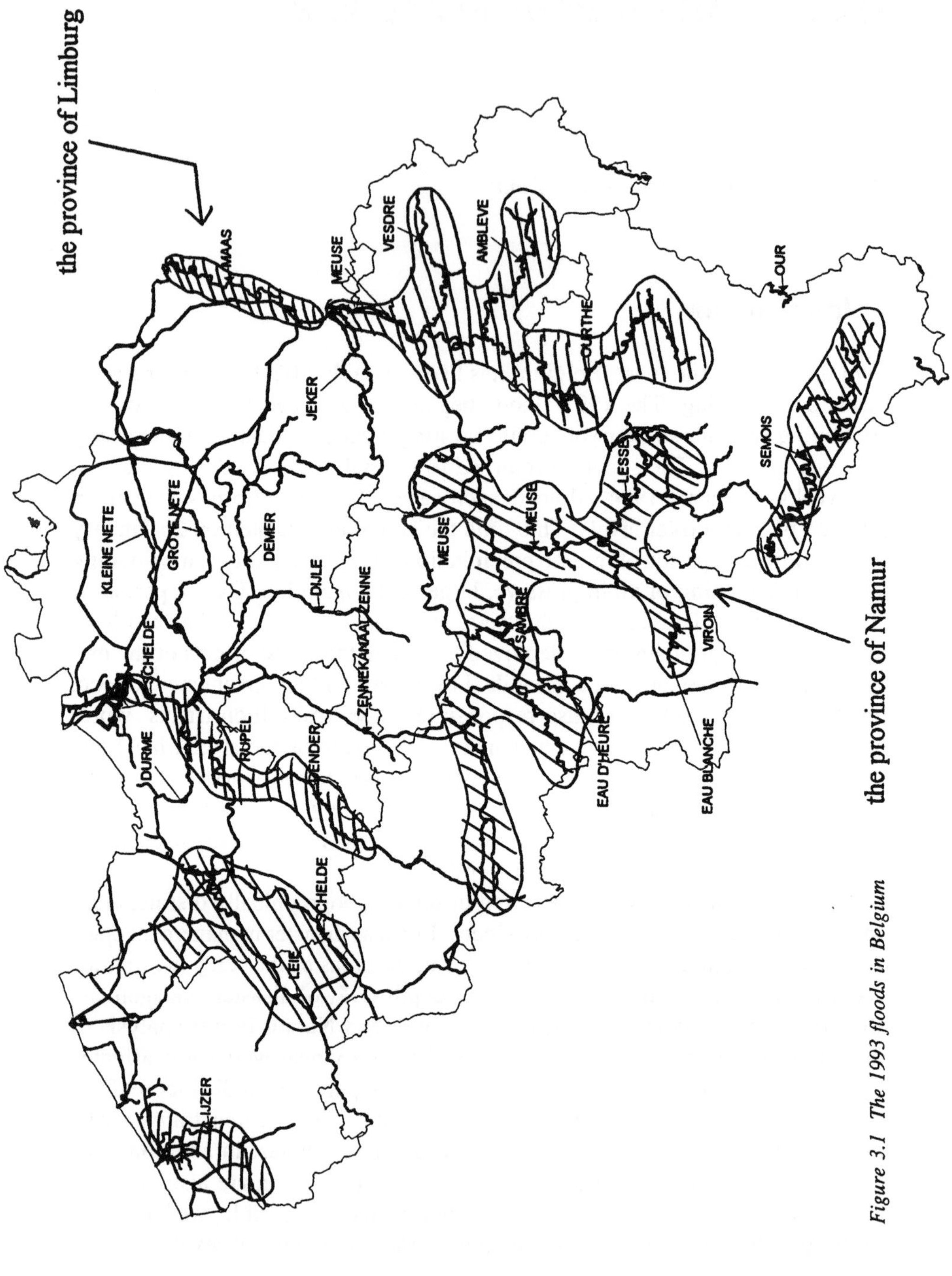

Figure 3.1 The 1993 floods in Belgium

Since both provinces were repeatedly affected by flooding of a similar intensity within a short space of time, it is possible to conduct a two-fold comparison in this study. A first comparison can be made between the crisis management as it was carried out in one province in two time periods, 1993 and 1995. The second compares the two provinces, taking into account the geographical, economic and cultural differences between the two regions.

2 The 1993 and 1995 riverine floods: an overview of events

The 1993 floods

At the end of 1993, many parts of Europe were affected by heavy rainfall. On 15 December 1993, persistent, heavy downpours over northern France, the Ardennes and Flanders led to an initial peak in water levels in many Belgian rivers. When these downpours continued over the next few days and water began to flow more rapidly off already saturated ground, river levels that were already high rose even further. This second peak coming on top of the first that caused the flooding.

The 1993 floods affected different, geographically separated areas (see figure 3.1). These included no less than seven provinces – four Wallonian and three Flemish provinces: Hainault, Namur, Liège, Luxembourg, East Flanders, West Flanders and Limburg. Only the provinces of Antwerp and Brabant escaped the danger of flooding. The scope of the floods can be explained by the wide geographical spread of the network of rivers in Belgium. The catchment area of the river Meuse covers practically the whole of Wallonia via the Sambre, the Viroin, the Lesse, the Semois, the Ourthe, the Amblève (Amel), the Vesder and countless smaller tributaries even before it enters the province of Limburg. In Belgium, the catchment area (i.e. the area covered by both the river and its tributaries) of the Meuse alone covers over 13,000 km^2. In the province of Namur, the greatest danger was posed by the Meuse, the Lesse, the Sambre, the Viroin and their many smaller tributaries. In Flanders, the situation was most critical in the province of Limburg. Here, the Meuse presented a constant threat of flooding for many municipalities and their surrounding villages over a length of 40 kilometres. A number of parishes were flooded while others were at constant risk of their dikes leaking.

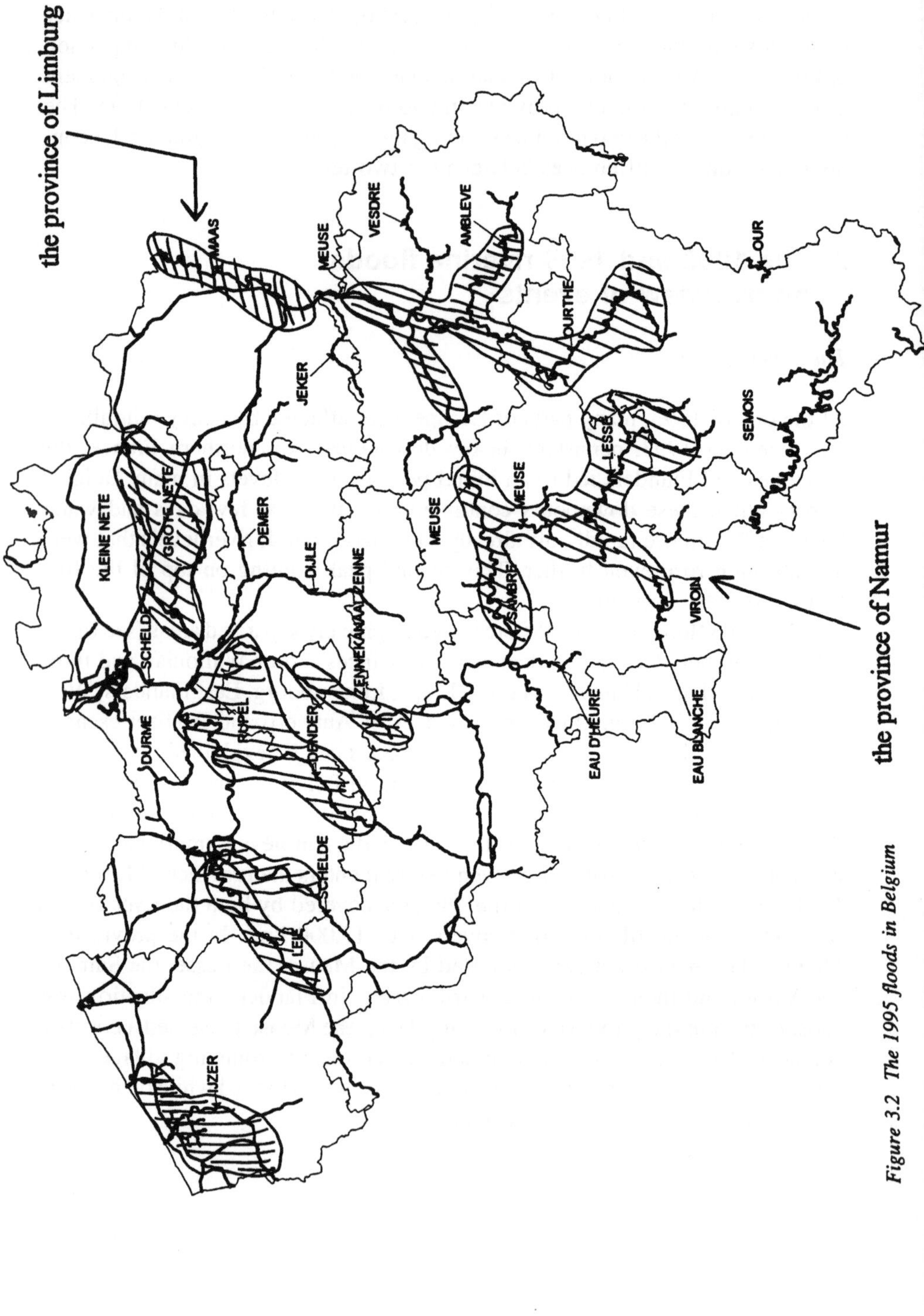

Figure 3.2 The 1995 floods in Belgium

The 1995 floods

On 22 December 1994 heavy rainfall caused a rise in the water levels on all rivers in the provinces of Luxemburg, Liège and Namur (see figure 3.2). The next day the water levels in the provinces of Hainault and Brabant were also rising to critical levels. On 28 December 1994 the authorities warned again for a rise in the water levels on all Wallonian rivers. At the end of 1994, there were also high water levels reported on the main Flemish rivers. On 2 January 1995 the situation returned to normal.

On 22 January 1995 the situation of almost all the Wallonian rivers worsened again. Over the short term, a fall in water levels was expected but persistent, heavy downpours over northern France, the Ardennes and Flanders led to a peak in water levels in many Belgian rivers. Due to this continuing rainfall the Wallonian and Flemish rivers rose to dangerous water levels on 26 January 1996.

The 1995 floods again affected different, geographically separated areas. These areas were spread across seven provinces: Hainault, Namur, Liège, Luxembourg, East Flanders, West Flanders and Limburg. In 1995 the provinces of Namur and Liège were most affected. The Meuse, the Lesse, the Ourthe, the Amblève and their many smaller tributaries caused serious flooding in several Wallonian villages and the inundation of various camping sites. In Flanders, the situation was most critical in the province of Limburg. During several days, the Meuse presented a constant threat of flooding for different municipalities. After a few days, the threat abated. High water levels started to cause problems in the province of West Flanders on the IJzer and the canal surrounding the city of Bruges.

3 Disaster management in Belgium

The structure of disaster planning

The development of a disaster management policy in Belgium has been a gradual process. Legislators tend to be particularly active following a series of cataclysmic events, but there is not always a forceful implementation. The specific events which prompted a further refinement of Belgian disaster management policy were the 1976 disaster at Seveso and the 1986 Chernobyl disaster. The consequences of the Chernobyl disaster, in particular, accelerated the planning process.

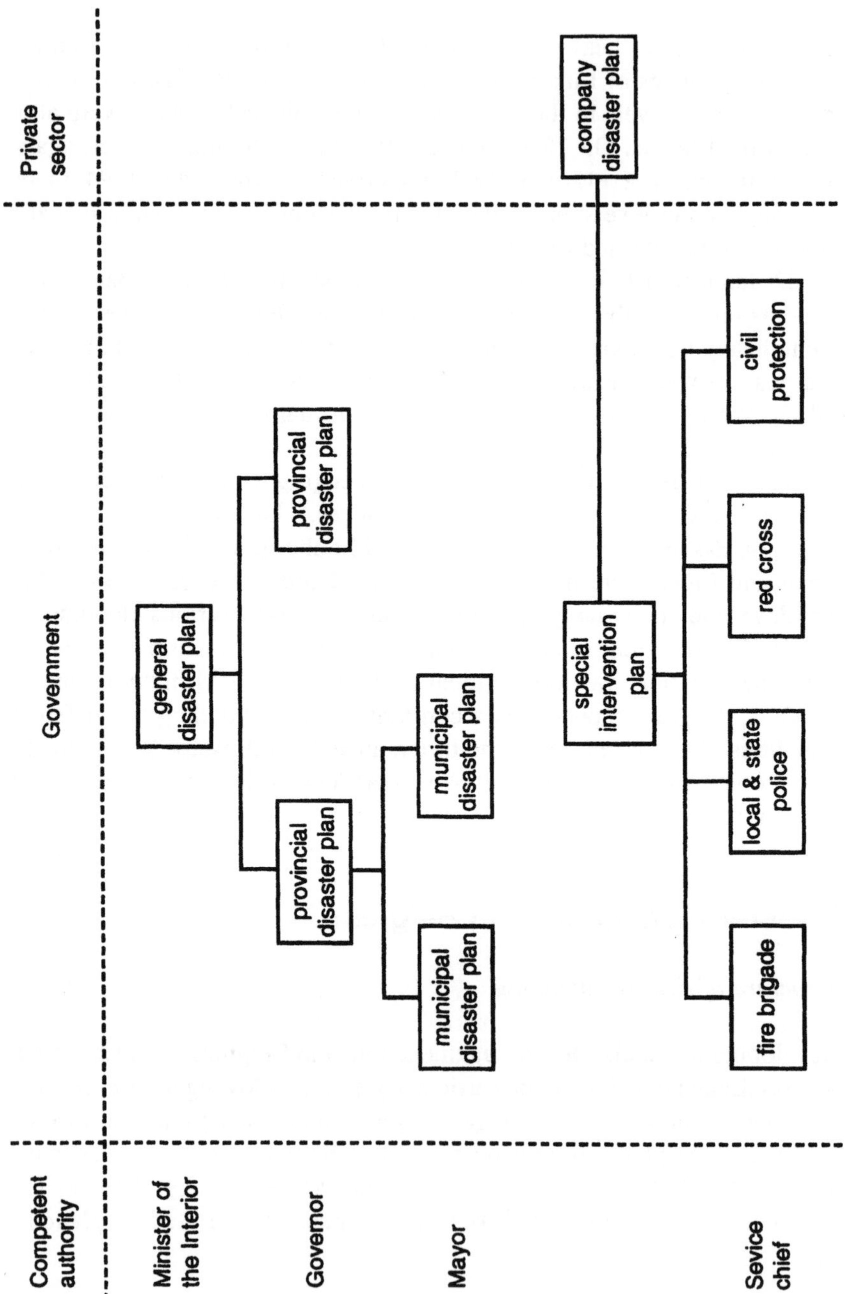

Figure 3.3 The structure of disaster planning in Belgium

At the federal level, responsibility for designing regulations for disaster planning clearly lies with the federal government and not with the provincial or municipal authorities. Throughout the various series of government reforms, responsibility for civil defence has remained a national matter (or rather a federal matter, following Belgium's change of status into a federal state). The Minister of the Interior coordinates emergency planning and mitigation. However, the federal government can only exercise its responsibilities effectively in consultation with provincial and municipal authorities since the regions are responsible for the environment and the communities are responsible for public health services. Thus, both provincial and municipal authorities play an important role in emergency planning and in providing practical support during disaster situations.

Emergency aid during disaster situations is organized on three administrative levels: municipal, provincial and federal. The mayor is responsible for safety in his or her own municipality. The mayor has authority over both the local police force and the fire service. At the mayor's request, the fire service must draft an intervention plan for those buildings and installations which pose a significant threat to the public in the event of a fire (obligatory for the so-named Seveso companies). Moreover, all operators of industrial enterprises covered by the Seveso Act must draft an internal company disaster management plan. This internal disaster plan must be tailored as closely as possible to the fire service's intervention plan. The mayor, together with the emergency services, can also draw up a municipal disaster plan. While the municipalities are not statutorily obliged to draw up such a plan, they are strongly encouraged to do so by the provincial and federal authorities. In contrast to the municipalities, the provincial authorities are obliged to draw up a general provincial disaster plan. This general plan must be supplemented by detailed practices in the event of specific disaster situations (obligatory for nuclear accidents and Seveso risks).

At the federal level, a national disaster plan for nuclear risks was drawn up in 1991. This is a framework plan in which the provincial disaster management plans and corporate emergency plans drawn up by the relevant institutions and the government services are outlined. This plan is activated by the Minister of the Interior who also announces the various alarm phases and heads the coordination committee. The Government Coordination and Crisis Centre, which was established in 1988, also plays a key role.[3] At the federal level, only a plan for nuclear disasters has been drawn up. At the time, there was no general federal disaster plan and no specific federal contingency plans.

Disaster planning processes

A 1990 decree brings together all of the legislation and regulations on disaster management planning in Belgium. The guidelines which have been formulated provide a framework for drafting a general disaster management plan. In the circular, disaster management plan is defined as:

> A plan setting out the broad outlines for action to be taken in the event of a disaster, catastrophe or serious accident, with a view to preparing government coordination of assistance by the various emergency services, such that all available and accessible material and personnel resources can be deployed in the most efficient way possible to protect the population and the environment against the negative effects of the situation.[4]

A general disaster management plan must be formulated at two levels: municipal and provincial. At the provincial level, the general disaster plan must be supplemented by specific contingency plans, since different accidents often require the application of specific operational management and safety techniques. Specific disaster plans should be drafted for each type of contingency, including road transport disasters, rail disasters, air accidents, flooding and other natural disasters, hazardous industrial accidents, accidents involving underground pipelines, serious oil spillage or interventions at sea. However, the provincial authorities are not obliged to draw up specific disaster management plans for all these risks. They must decide which risks are most significant within their own territories. They are only legally obliged to draw up contingency plans for so-called 'Seveso' or nuclear establishments in their province.[5]

The provincial governor is appointed crisis manager in disaster situations which affect more than one municipality. He is responsible for drafting a general provincial disaster management plan. The province should also draft a model municipal disaster management plan, designed to make the municipal plan fit the provincial plan. The model plan must be converted by the municipalities into a general municipal disaster management plan. The municipal disaster management plan is activated by the mayor after he or she has been notified of the disaster situation by the emergency services. In particularly urgent cases, this decision can also be made by the chief fire officer, the local police chief or the Gendarmerie (national police, in Flanders known as the Rijkswacht) commander. The mayor notifies the provincial governor and informs him about the circumstances of the disaster, the risks involved and how the situation is developing. The provincial disaster plan is activated by the governor following consultations with the mayor. In principle, it is activated whenever different emergency services need to work together to deal with a particular event. Indicators of the seriousness of a disaster include the number of injuries or fatalities and the presence or release of hazardous substances and gases which constitute a threat to the population.

Phasing of disaster management

The disaster plan unfolds in four phases, depending on the size of the incident, its anticipated effects and the type of emergency aid required (see table 3.1).

Table 3.1 The organization of emergency management in Belgium

Phase	Level of Administration	Responsible Authority
1	Municipal level: limited action and coordination	Mayor
2	Municipal level: upgraded emergency action and coordination	Mayor + municipal crisis centre
3	Provincial coordination	Provincial governor + provincial crisis centre
4	Federal coordination	Minister of the Interior + government crisis and coordination centre

These phases are not necessarily implemented in chronological order. In order to be able to respond efficiently to a disaster situation, each municipality should have its own disaster management plan. Moreover, there must be continuity between the municipal plan and the provincial plan, and between the coordination carried out by the mayor and the subsequent coordination carried out by the provincial governor. The municipal plans should therefore be drafted according to a standard format, precisely in order to preserve this continuity of emergency operations.

The decree on disaster planning of 19 June 1990 specifies that disaster management plans must at least contain the following information: alarm procedures, action procedures for the various emergency services and organizations, preparation of individual protection measures, the coordination of the operations and the procedure for notifying those affected by the incident. The main provisions are discussed below.

The alarm procedure — In 1988, a special obligation was introduced to notify the authorities in the event of a serious accident linked to certain industrial activities which has since been applied to all kinds of emergencies.[6] First, there is the obligation to dial the emergency number (100), which will launch the emergency services. Second, the Government Coordination and Crisis Centre must be notified. Standard notification forms are issued to ensure that the notification procedure is carried out as uniformly as possible. The methods used to notify the population (sirens, radio or TV bulletins, cars with loudspeakers) must be comprehensively covered in every disaster management plan for emergency aid provision for the 'Seveso' companies. The population can also be instructed to take certain measures through the public service broadcasting stations (BRTN, RTBF and BRF). A standard media bulletin for notifying the population was also devised.

Division of labour and coordination — In a disaster management plan six functional disciplines are discerned. For each task, an indication is given of who the on-site commander is, who on the coordination committee represents it, which resources are available, and which action guideliness apply. The various branches are:

1 emergency aid operations (fire service);
2 medical and sanitary aid (the provincial health inspector, assisted by the Red Cross and hospitals in the area);
3 policing affected or threatened areas (local police forces and/or Gendarmerie);
4 various activities, relief aid, transport and logistics (civil defence force);
5 informing the population in the affected area (mayor or provincial governor);
6 judicial and technical investigations (public prosecutor).

During phases 1 and 2 of the disaster management plan, the mayor is responsible for the provision of emergency aid. He can form a *municipal coordination or crisis committee* to support him in this. The main task of this coordination committee is to assist the mayor in evaluating the disaster situation, to decide on the deployment of the necessary resources, to help select the methods to be used and to investigate the problems which have arisen. All its members must ensure that the committee's decisions are carried out promptly and correctly by the municipal services. This municipal coordination committee consists of the mayor (chairperson), the aldermen, the head of the local Social Welfare Office, the head of the municipal technical service, other relevant city officials, the local fire chief, the local police commissioner, and representatives of the Gendarmerie, the civil defence force and the Red Cross.

The provincial governor assumes overall authority during phase 3 of the disaster plan. The governor is also responsible for appointing a *provincial coordination committee* to help him administer the disaster management operation.

This provincial coordination committee consists of the governor (chairman), his chief political adviser and the relevant district commissioner, the mayor of the municipality involved, the provincial head of the civil defence force, the military commander of the province, and representatives of the Red Cross and the Gendarmerie. During phase 3 of the disaster plan, the municipal coordination committee continues to administer the operations of the municipal services, offers any help requested by the provincial coordination committee and carries out any decisions made by this committee. It is advisable for the mayor of the municipality involved to be a member of the provincial coordination committee since this will facilitate inter-governmental communications.

Apart from the establishment of the coordination committee (which is the highest administrative authority in a disaster situation), a operational command post is set up immediately in the disaster area. The fire chief coordinates the emergency aid and rescue operations in the disaster area. He does this in accordance with the instructions issued by the mayor or the coordination committee. The fire chief and the heads of the various emergency services will then constitute the operational staff of the command post. The heads of the emergency services nevertheless remain fully responsible for the performance of their own organizations. More than one command post can be set up, depending on the nature and extent of the disaster area, for example if several municipalities are involved. Each command post is headed by the local fire chief and is answerable to the provincial coordination committee.

Informing the public — During phases 1 and 2, all information supplied to the press is provided solely by the mayor himself (or his or her deputy). From phase 3 on, the provincial governor is responsible for providing the necessary information to the public. In the event of serious disasters, the governor must also open an information centre for victims or their families where the media can also obtain information about the emergency operation. This is required when injuries and/or fatalities have occurred or when dangerous gases and products have been released which constitute a threat to public health. A special telephone number can also be established for this purpose. In addition, the mayor and governor can appoint a spokesperson whose task is to direct and substantiate the flow of information to the press and the public.

Emergency management in the Limburg and Namur provinces

Since the 1970s, several very different disaster management plans have been drafted in many provinces. These plans were often tailored to specific risks and at the time, no general disaster management plan or uniform approach to disaster management existed. Such uniform disaster management plans at all policy levels were only given any definite encouragement following the guidelines issued by the federal government in 1990. Given the focus of this chapter, we shall briefly describe the outlines of disaster planning in the provinces of Limburg and Namur.

Limburg — The province of Limburg has always taken the lead in Belgian disaster management. It first issued a general emergency plan in 1979 and since then, the plan has been continuously refined. In 1993, a new revised and extended emergency plan for Limburg was designed based on the statutory guidelines issued in 1990.[7] All of the adjustments and refinements which had been made over the past 14 years were included in the new plan.

The General Emergency Plan for Limburg consists of two large volumes, the first of which sets out the provincial coordination system for all types of serious accidents or disasters. The second part brings together the main details of the individual municipal emergency plans. The Provincial Coordination Plan consists of 11 chapters. The second chapter forms the core of the plan. It contains a comprehensive description of the actual operation of the emergency plan. It covers in detail the alarm procedure, the phases of the disaster management, the coordination process, the six key disaster management functions and the communications plan.

The second part contains information and documents concerning disaster management at municipal level. It briefly discusses the powers and tasks of the municipalities and gives the provincial model for municipal disaster management plans. The third chapter synthesizes the main features of the individual emergency plans of the 44 municipalities in Limburg. This information is ordered in accordance with the provincial model for municipal disaster management plans.

Namur — In the province of Namur, a general plan for emergency action was drafted in 1988.[8] This disaster management plan developed gradually and by 1995, it largely reflected the principles and procedures set out in the 1990 statutory guidelines. The general plan for emergency action remained almost totally unchanged following the 1993 and 1995 floods. Nevertheless, these emergency operations did have an effect on provincial and municipal disaster planning in Namur. Currently, efforts are being made to draft a specific disaster management plan to cope with flooding. This plan is being prepared by the provincial executive in close cooperation with other competent authorities and with the relevant emergency services. Following the experience gained in the 1993 and 1995 floods, the provincial executive of Namur has also compiled a provincial model for municipal disaster plans. Unlike the municipalities in Limburg, most municipalities in the province of Namur did not have their own disaster management plans. The provincial model is structured along similar lines to the general provincial disaster plan, except that it focuses specifically on disaster policy at local level. The provincial executive is hoping this will encourage the municipalities to devise their own municipal disaster plans.

4 Risk communication

The warning process

The organization of warning — The hydrological services of the Flemish and Wallonian Regions are responsible for identifying potential problem situations on and around Belgian rivers. The authority responsible for the Flemish rivers is the Hydrological Survey Department of the Ministry of Environment and Infrastructure of the Flemish administration. In Wallonia, it is the *Service d'Etudes Hydrologiques* (Hydrological Survey) of the Wallonian Ministry of Transportation and Infrastructure.

These services monitor the situation on Belgian rivers on a daily basis by means of a network of limnigraphs or measurement stations situated at various points along the rivers. These stations measure the water level via a system of plotters and floats. The measurements obtained are then usually transmitted to the computer room of the hydrological survey twice a day, and are subsequently processed. In the event of a problem situation, the frequency of these readings increases substantially, sometimes to 30-minute intervals. These services also have access to a network of rainfall meters which indicate how much precipitation has fallen in a specific area. Finally, they are also connected by modem to the Royal Meteorological Institute, so that they are kept up to date on the latest weather forecasts.

On the basis of this data, the hydrological services can identify when certain rivers are in danger of flooding. When a specific water level has been exceeded, an early warning phase is activated, and when a critical limit has been exceeded, the emergency alert phase is activated. In such situations, a monitoring unit is set-up within the hydrological service to monitor developments closely and is responsible for supplying information to the various groups involved. These groups are, first, the federal government's crisis and coordination centre which in turn sends information to authorities in the threatened area, including the provincial governor, the mayors of the alerted municipalities, the civil defence force, the fire service, and the Gendarmerie. The hydrological services also inform agencies responsible for the management of the rivers and dikes, such as the Albert Canal Service. The Ministers of Public Works in both regions are also notified of the situation. Contact is also made with the Department of Public Works in the Dutch province of Limburg, which compiles and supplies forecasts of expected water levels.

A second important actor with responsibility for forecasting potential disaster situations is the Royal Meteorological Institute (KMI). The main task of the KMI is to provide the public with the meteorological and climatological data and information they require. During extreme weather conditions, this institute plays an important role in providing up to date climatological data and in forecasting further meteorological developments. When forecasting exceptionally severe weather (violent thunderstorms, storms accompanied by squalls, heavy precipitation), the KMI issues 'Special Flashes' several times a day in addition to its standard bulletins. These 'Special Flashes' are faxed to news organizations such as the BRTN, RTBF and RTL-TV1 as well as to the government's crisis centre, the fire service, the Gendarmerie and the civil defence force. If requested, the KMI also sends direct bulletins to those in charge of directing local disaster operations.

The warning process in 1993 — On 13 December 1993, the hydrological services (Sethy) issued an early warning to the federal government's crisis and coordination centre. This message warned of the danger of local flooding in the basins of the Rivers Semois, Viroin and Lesse in Wallonia. The crisis centre notified the relevant authorities and emergency services in the provinces of Namur and Luxembourg of the situation. On 14 December, the early warning notification was extended to a wider area in Wallonia. During the next few days, the Sethy reported an improvement in the situation in the Wallonian rivers, but upheld the early warning for the Upper Meuse and Semois.[9]

On 19 December, the situation deteriorated to such an extent that the Sethy issued a general early warning for all of the rivers in Wallonia. The Sethy immediately sent a bulletin to the federal crisis centre, which in turn notified the relevant provincial executives and emergency services of the situation. The alarm procedure was also announced for one or two specific problem areas. In these areas, the emergency limit had been reached and there was a danger of local flooding, especially in the many camping sites along the flood plains. In the late afternoon, the crisis centre decided to warn the caravan owners by means of announcements in the media. From 20.00 onward, the RTBF and BRTN radio news bulletins carried warnings of high water levels and their likely effects for caravans parked in the winter flood plains of the Semois, Vierre and Rulles.[10] The owners of these caravans were requested to move them as quickly as possible. It was also announced that there was a danger of flooding on other rivers, including the Sambre, Zenne and Dender, but only where these rivers flowed through Wallonia. The weather bulletin forecasted continued rain which was expected to lead to a risk of flooding, especially in the Meuse Valley. Until this time, the authorities still believed the problem would be limited to local flooding and to the inundation of camping sites in the winter flood plains of these rivers.

On 20 December 1993, the situation on the Wallonian rivers escalated rapidly and severely. In the early afternoon, the Sethy sent continual bulletins confirming a general rise in all rivers. Moreover, at around 9.00, the KMI announced in its first 'Special Flash' that heavy rain was expected to continue to fall over the next few hours. At around 12.00, the government's crisis centre sent a fax to all provincial governors in Wallonia asking them to make preparations to activate provincial coordination operations. At around midday, the Sethy issued a general alarm for all Wallonian rivers. In view of the fact that a further rise in the water level was expected, flooding was forecast in many places by the end of the day and during the night. This information was confirmed by a second 'Special Flash' from the KMI. Throughout the afternoon, based on the local developments of river levels, each of the four Wallonian provinces announced in turn that they were launching phase 3 of the disaster management plan. In the province of Namur, the provincial coordination committee was set up at around 14.00 to head the subsequent emergency aid operations.

Because high water levels were expected mainly on the River Meuse and its tributaries, the federal crisis centre also notified the governor of the province of Limburg in Flanders at 13.00 on 20 December. On 20 December, at around 12.00, the responsible official at the Flemish Hydrological Survey Department sent an initial fax to the federal crisis centre, in which he warned of 'critical water levels leading to a possible danger of flooding on certain Flemish rivers.' On the same Monday morning, the River Meuse in Limburg reached the emergency alert level of 46 TAW at the Smeermaas measuring station.[11] During the morning of 21 December, however, it quickly became clear that the emergency alert level would be substantially exceeded. At 10.45, phase 1 of the provincial disaster plan was therefore announced. The situation escalated so rapidly that by 11.30, phase 3 had to be launched. The various emergency services were immediately notified and called to action. The provincial coordination centre began operating in the fire service headquarters at Maasmechelen almost immediately. At 12.30, the first crisis meeting of the provincial coordination committee was called under the chairmanship of the provincial governor. Throughout the subsequent emergency aid operations, both hydrological services and the KMI continued to send updates (at more regular intervals) to the federal crisis centre and to the provincial coordination committees. These updates included the latest reports on the situation on the various rivers and the rainfall situation. Wherever possible, they also gave updates on the expected rainfall and further changes in water levels.

The 1993 floods took the relevant organizations, authorities and the various emergency aid services completely by surprise. The extent of the disaster situation was unexpected and unforeseen. This can be partially explained by the sudden way in which the flooding occurred. Yet it cannot be denied that there were also one or two factors which should have led to a greater alertness and the taking of necessary precautions. The early warning notification had been in force in Wallonia since 13 December 1993. However, the continued rainfall meant that the soil was completely saturated and could not hold any more water. Any further rainfall would have to be collected and drained off by the rivers. Consequently, forecasts of further rainfall should have been a signal that problems were to be expected.

The hydrological services are initially responsible for communicating the latest water levels. By means of an extensive network, these services can assemble an adequate amount of information concerning the situation on the rivers. For the provincial crisis committee, this information was indispensable to obtain an accurate picture of the level and development of water levels. Such information was important in enabling them to make timely decisions relating to evacuations, the deployment of manpower for emergency interventions, and so on.

The hydrological services were sent their data regarding water levels "online", after which they were included in a bulletin and sent by fax to the federal crisis centre. The crisis centre, in turn, faxed the bulletin to the provincial coordination committees and the relevant emergency services. In practice, this meant a delay of one-and-a-half to two hours between the actual measurement of the water levels and the receipt of this information by the emergency aid workers in the disaster area. In the province of Limburg, it meant that by this time, the emergency aid workers operating in the disaster area had usually assessed the situation for themselves, and only later had this assessment confirmed by the report from the hydrological services. The measurements and bulletins by the hydrological services were therefore important chiefly during the earliest warning and emergency alert phases. They were also indispensable in enabling an accurate forecast to be made of the development of water levels and flow rates. This was necessary before making (often technical) decisions such as whether to strengthen weakened and saturated dikes, and the positioning of additional pumps. In the province of Namur, on the other hand, it was vital that the provincial coordination committee was regularly briefed about the situation on the various rivers. Without this information, it would have been impossible to obtain an accurate picture of the situation in the disaster area.

The hydrological services were less well-equipped to make accurate predictions (for more than 12 hours. They had to rely primarily on the Sethy and on the Dutch Department of Public Works for predictions about how water levels on the River Meuse were likely to develop. Although the Sethy had access to a number of forecasting and simulation models, it was not entirely able to forecast water levels across an extensive network of rivers. Moreover, these services were also heavily dependent for their forecasts on the evolution of water levels in neighbouring countries and on the information that was available concerning these non-Belgian rivers.

Although the flow of information between the two language regions went fairly smoothly, deficiencies were nevertheless recorded on the Flemish side. For example, the provincial coordination committee in Limburg could not always obtain precise information about the water level situation upstream in Wallonia. This flow of information was often late and did not always reach the coordination committee directly.

Contacts and the exchange of information with the relevant authorities abroad were laborious and prone to delays. During the early phases, there was no information from France concerning water levels on the Meuse, Scheldt or IJzer or their tributaries, even though this could have improved the predictability of the problem situation. Information was exchanged with the Dutch Department of Public Works only once the actual problem situation had developed.[12]

The warning process in 1995 — On 22 December 1994, the Sethy sent an early warning bulletin to the federal government's crisis and coordination centre. This bulletin warned of high water levels on all rivers in the provinces of Luxembourg, Liège and Namur. The next day, the early warning was extended to the provinces of Hainault and Brabant. After several hours, the situation stabilized. On 28 December 1994, an alarm was sounded for the Senne and the Samme in the province of Namur. The Sethy also warned of a rise in the water levels on all Wallonian rivers. The federal crisis centre then sent a press release to the public broadcasting authorities and the national Belgian press agency Belga. On 28 December 1994, a warning of a rise in water levels on the Wallonian rivers was carried for the first time by the media. Caravan owners were asked to move their property as quickly as possible. The next day, another general early warning was issued for all of the Wallonian rivers. This was retracted on 2 January 1995. There were problems on the River Ourthe in the provinces of Liège and Luxembourg on 11 and 12 January 1995, but from 13 January the situation returned to normal. At the end of 1994, the Flemish Hydrological Survey (Diho) also reported continued high water levels on the main Flemish rivers.

In 'Special Flashes' which it issued on 21 and 22 January 1995, the KMI warned of continued rainfall. On 22 January the situation worsened, prompting the Sethy to again issue a general early warning for all the Wallonian rivers. On 22 January 1995, BRTN radio was the first to give a comprehensive report regarding the flooding in various Wallonian villages and the inundation of various camping sites. On 23 January 1995, the crisis centre again sent out a press bulletin indicating the locations of possible problem areas. It warned that in many places, the emergency alert level would be surpassed over the next few hours and that flooding could occur. From this time on, the flooding in Wallonia and later also in Flanders became one of the main items in the news bulletins of the various media organizations.

During the night of the 22 January, the situation on the Wallonian rivers became quite serious and on 23 January at around 9.00, the Sethy sounded a general alarm for all of the Wallonian rivers. Due to continuing rainfall, a general rise in water levels and flow rates was predicted over the next 24 hours. On 23 January at around 14.00, the Flemish Hydrological Survey Department issued a bulletin which warned of high water levels and rapid flow rates on the Flemish rivers. With the exception of the Meuse, which was expected to continue to rise, a stabilization in the situation was expected during the course of the day. On 24 January, the KMI again predicted continued heavy rainfall. On that day, the Meuse reached 46.46 TAW at the Smeermaas measuring station. Over the short term, a fall in water levels was expected, but from 26 January the level of the Meuse again rose to dangerous levels. From that moment on, it was decided that the provincial coordination committees would take over the leadership of the emergency aid operations. Throughout the remainder of the emergency operation, both hydrological services and the KMI continued to send updates several times a day to the federal crisis centre and the provincial coordination committees.

This time, the various bulletins about the early warning situation on the rivers were heeded. Both the authorities and the different emergency aid services had taken the necessary precautions well ahead of time. From 22 January on, the emergency aid services were already active in many high risk areas and the necessary preparatory measures were taken. The authorities were also following developments closely and responded rapidly. In the Wallonian provinces, phase 3 was announced on 26 January and the provincial coordination committees launched their operations. In the province of Limburg, there was an initial meeting of the provincial coordination committee on 22 January in which the necessary preparations were made. On 25 January, this was followed by an evening crisis meeting and on 26 January the provincial governor decided that phase 3 of the disaster management plan should be announced. The relevant emergency services responded rapidly and effectively.

The alarm procedure which was activated following the 1995 floods went smoothly and without delay. Before the situation escalated on 26 January 1995, the hydrological services and the KMI had issued an early warning several times. The actual danger was estimated in good time. The various services were brought to a state of readiness and the necessary preparatory measures were arranged. As a result, those involved were already prepared to respond at the moment when the water struck.

In 1995, the exchange of information with the hydrological services in the countries bordering Belgium was carried out promptly and satisfactorily throughout the emergency aid operation. However, the bulletins from the hydrological services were often couched in highly technical and unclear language. As a result, they could only be read by experts and caused bewilderment among authorities and emergency aid workers. Moreover, the hydrological services in the neighbouring countries often used different scales of measurement for assessing water levels.[13] This sometimes led to confusion and faulty interpretations.

The experience gained in 1993 clearly taught those involved how quickly such a situation can escalate following prolonged periods of rainfall. In 1995, the actual danger was therefore assessed well in advance and the various emergency services were brought to a state of readiness and the necessary precautions were taken. Because the various services responded in time and the widespread damage that followed the 1993 floods was prevented.

Warning and informing the public

During phase 3, the provincial governor and the provincial coordination committee are the highest responsible authorities. As such, they are ultimately responsible for issuing warnings and informing affected or at-risk populations. This does not prevent other policy bodies from being active in this area. For example, in various provincial disaster plans, responsibility for notifying the population is delegated to local mayors who are then specifically responsible for informing their own local populations. In cases where the disaster situation is more severe and extensive, it is also possible for federal bodies to take over. The federal government's crisis and coordination centre plays a key role here since many services and organizations report daily to this centre about the situation in many different sectors and about possible problem situations. If a situation escalates rapidly, the Minister of the Interior can decide to notify the population through the media. To this end, unofficial agreements have been reached between the government and the public broadcasting authorities. There is no legal provision obliging public broadcast agencies to transmit government information bulletins following crisis situations. However, cooperation agreements to thyis effect have already been concluded between the BRTN and the governors of the Flemish provinces and between the RTBF and the Wallonian provinces. These agreements are based on a standardized bulletin containing information about a given disaster situation. Agreements have also been reached between the BRTN and RTBF and the federal government, but there are as yet no such agreements with the municipalities.

Public information in 1993 — In 1993, the rise in water levels in the Meuse and its tributaries was sudden and unexpected. The size and speed of the 1993 floods took both the authorities and the various emergency aid services by surprise. No early warning was issued to the population directly at risk, and these communities were usually left to determine for themselves how seriously the situation was deteriorating. On 19 December 1993, however, there was a press announcement from the federal crisis centre. This bulletin warned the owners of caravans parked on the banks of a number of Wallonian rivers of the danger that the camping sites might be flooded. The selected radio news bulletins also warned of heavy rainfall over the coming hours and that this might lead to flooding in Wallonia.

In the provinces of Namur and Limburg, the public was kept informed largely through the media (radio, TV, newspapers). There was almost no additional means of spreading information, such as information leaflets, folders or cars with loudspeakers. In Limburg, specific agreements were made within the coordination committee with regard to the provision of information. Local mayors were delegated to keep their local populations informed of developments. Because these mayors were members of the crisis committee, the flow of information to the affected populations was largely sent through the provincial coordination centre. In Namur, however, the geographical scope of the disaster areas was much greater. As a result, the distance between the provincial coordination committee and the affected populations was also greater. In the smaller Wallonian villages in particular, the mayors and the municipal services played a crucial role in informing the public about the rescue operations and the local situation. Few additional means of information were provided, and consequently, news was largely distributed by word of mouth.

In 1993, the population in high risk areas was not informed in advance of the danger of flooding. Only caravan owners were warned, and then only at a late stage. As a result, almost no precautionary measures were taken by the population. The affected populations were also not notified in advance of the intention to evacuate people in Limburg (some 2,000) or in Namur (some 500). The Red Cross had urged the authorities to distribute a leaflet with a list of instructions to the populations at risk. The provincial coordination committee decided not to act on this advice, fearing such a leaflet might result in panic. The actual result was that people ultimately had to be evacuated in an unexpected an unprepared state, and causing many problems and inconvenience for both the affected families and the emergency services.

The process of notifying the population was not always properly coordinated or managed. In Limburg, the various mayors each acted as independent spokesmen. There was little consultation about what information would be communicated to the public. Moreover, it is quite likely that each mayor was at the time trying to raise the profile of his or her municipality to the media, which would have the effect of minimizing the problems and needs of the other municipalities. In Namur, the provincial coordination committee was responsible for distributing information to the media.

Public information in 1995 — During the December 1994, the authorities, emergency services and the populations at risk were warned several times of a possible rise in water levels on many rivers. The situation was closely monitored by the various media organizations. Before the situation escalated to a disaster on 26 January 1995, the population had been warned repeatedly and in time about the possible danger of flooding. They were also urged to take the necessary precautions. The federal crisis centre sent several press releases to the public broadcasting authorities and the Belga press agency. On 28 December 1994, the population was warned a first time of the rise in water levels on the Wallonian rivers. Caravan owners were asked to move their property as quickly as possible. The local flooding in Wallonia was not yet a major news item at this time.

During the first week of January 1995, the situation on the rivers stabilized, making additional public service broadcasts unnecessary. Starting 22 January, the emergency services (fire service and civil defence force) began their first interventions in response to the local flooding. In some areas in the valley of the River Meuse, the situation deteriorated gradually, prompting a number of municipalities to activate phase 1 of the disaster plan (increasing emergency intervention activities). In certain municipalities, municipal crisis centres were opened (phase 2 of the disaster plan) to facilitate the coordination of local operations.

On 23 January 1995, the federal crisis centre again sent out a press release indicating the locations of possible problem areas. It warned that in many places, the emergency alert level would be surpassed over the next few hours and that flooding could occur. From this time on, the flooding in Wallonia became one of the main items in the news bulletins in the media. By the time the situation eventually escalated on 26 January 1995, most people had already taken the necessary preventive measures. Many people had taken their valuable belongings to the top floors of their houses. They had also placed sandbags and taken the necessary safety precautions. Farmers took the necessary steps to protect their livestock.

The subsequent provision of information about emergency aid operations again, was done almost entirely through bulletins via the media (radio, TV, newspapers). The provincial heads of policy organized press conferences several times a day to this end. This had a positive effect on the distribution of information in general, as well as on the provision of information to the populations at risk. On 31 January, the Minister of the Interior also held a press conference to discuss the processing of claims following the floods.

In Limburg, the affected populations were informed prior to evacuation. A list of instructions was distributed to all of the inhabitants at risk. In Namur, the evacuations following the 1995 floods were carried out without any prior warning to the affected populations. This undoubtedly complicated these operations. Moreover, a timely decision coupled with a warning of the population could have helped to minimize public resistance to the evacuation.

Media management

The procedure for notifying the public and the media in the event of a disaster is contained in the provincial disaster plans. The designated authority responsible for supplying, monitoring and coordinating the flow of information depends on the phase involved. In phases 1 and 2 it is the mayor, while in phase 3 it is the provincial governor or his deputy (the district commissioner) and in phase 4 it is the Minister of the Interior. Each of these individuals can also appoint a specific external spokesperson.

Media management in 1993 — Specific practical agreements were made within the coordination committees in Limburg and Namur with regard to the provision of information and public notification. The governor took on the task of providing information about the general situation on the Meuse and the progress being made by the emergency operations. The mayors were given the task of informing the local populations. The heads of the emergency services present were responsible for providing information about their own services, while the engineers employed by the Department of Public Works were responsible for providing technical explanations about the state of the dikes. No external spokesperson was thus appointed by the provincial coordination committee, even though the district commissioner had argued strongly for this to be done in the interests of an efficient and accurate supply of information.

In Limburg, a large number of media representatives were present in the disaster area. This concentration of the media was possible because the disaster area in Limburg was clearly defined. They reached the scene only a few hours after the activation of phase 3 in the immediate vicinity of the coordination committee's headquarters at the fire station in Maaseik. They included the BRTN (radio and television), the commercial television channel VTM, local radio channels and journalists representing various local, regional and national newspapers. These media representatives would remain in the area almost constantly over the next few days. Two press conferences were held each day. Visits of various ministers to the affected municipalities were also frequently used to organize additional press conferences. For any further interviews or information, the journalists were able to approach the coordination committee or the headquarters of the Red Cross, where a telephone number for use by the public or the press was permanently available.

In Namur, as in the other Wallonian provinces, the disaster area was much larger and more extensively spread. As a result, the press were not concentrated in a single location but widely distributed among the many affected areas. The central reception point for the journalists was in the provincial headquarters in the city of Namur, where press conferences were organized in response to developments as they occurred. For additional information or an outline of the latest situation, the press relied on information from the mayors, on witnesses from the local populations and on emergency aid workers on the spot.

Public broadcast agencies fulfilled a dual role during this crisis situation. By transmitting official public service bulletins, they were partly responsible for keeping the affected populations informed. They also provided updates on the disaster situation and on the emergency operations by means of news reports. However, both roles were kept strictly separate. The official public service bulletins were given a special slot in the news reports, for example, by the fact that the announcer stated that a public service bulletin was about to follow. The official information supplied by the coordination committee to the media was generally provided in a timely manner and in good form. The various appeals to the population by the coordination committee were promptly broadcast on radio and television, however, a number of inaccurate bulletins and exaggerated accounts were circulated by the media.[14]

At first sight, the influence of the media bulletins on the decision making process regarding emergency intervention was not great. The nature of the disaster situation was such that there was little opportunity to comment on the way the resources and manpower were deployed. Yet, in view of the confused nature of the emergency operation and the fairly weak communications policy pursued by the provincial authorities, it is surprising that there was not more criticism by the media. The news reports tended instead to concentrate in the first place on the dramatic nature of the disaster. They were complimentary about the interventions by the emergency services and the decision to deploy hundreds of volunteers, sometimes even describing these efforts in heroic terms. There was little criticism of the rescue operations. Criticism about the causes of the floods and the approach to dealing with them only began to emerge well after the operations had ended.

One area of policy was significantly influenced by media reports, and this was in the recognition of the emergency as a full-blown disaster. This recognition was important in that it allowed victims to apply for compensation under the federal disaster fund. Since private insurers were reluctant to compensate damage caused by the flood water, such official recognition was the only possible way in which these victims were likely to obtain financial reparation. The nature of the news broadcasts (i.e. placing the victims at the centre of the reports) and their regularity meant that the federal authorities were, to some extent, pressured to make a positive decision. As early as the third day of the emergency operations, the Minister of the Interior announced that the event was a disaster and that victims were eligible to claim compensation from the national disaster fund.

Information policy was not given enough practical and material support. No separate room was set aside for the press. Similarly, no facilities were provided for the press to set up their vehicles, broadcasting equipment and other apparatus. The press conferences were organized on a somewhat improvised basis. The press conferences themselves were held at irregular times and during the first few days, no written information was provided for the journalists. Due to the overloading of telephone lines at the crisis committee headquarters, it was difficult for journalists to reach any of the officials responsible in between press conferences.

Public information duties were too fragmented among the various members of the crisis committee, whose actual core tasks at that time clearly required them to be elsewhere. The local mayors were keen to act as spokespersons, but the various bulletins they gave were often insufficiently coordinated. A number of inaccurate bulletins were circulated.

Media management in 1995 — In 1995, the floods and the emergency operations in the various provinces were again the main focus of interest in national, regional and local news reports. The huge and varied media presence was managed much better this time. In Namur and Limburg, the heads of the provincial crisis centres held a press conference twice a day at appointed times. The contents of these press conferences were prepared in advance and the journalists were usually able to use prepared press releases or additional documentation. Professional spokespersons were appointed. A press liaison was set up where journalists could go at any time with their questions.

In 1995, the Minister of the Interior also organized a press conference in collaboration with the federal government coordination and crisis committee. This press conference was only open to the national media. The minister briefly described the disaster situation in the various provinces and gave a broad overview of the emergency interventions. The main item on the agenda was the government's official recognition of the event as a disaster and the initiation of the disaster fund. The minister announced a series of measures intended to process the claims more rapidly. This was especially important since the 1993 claims had not yet been paid, and there had been some media criticism of the slow processing and payment of these claims. These claims were now joined by hundreds of additional claims for the 1995 floods which had further slowed down the process.

The press conferences held in the provincial crisis centres were attended primarily by the regional and local media. The national media tended to go instead to the federal crisis centre for information. In Namur in particular, local media played an important role in keeping the affected populations informed, and the role of the local radio stations was especially crucial. In Wallonia, the network of local radio stations was much larger than in Flanders. In Limburg, the new regional television station 'TV Limburg' also played a major role in keeping the affected populations informed.

The appointment of spokespersons clearly had a positive effect on the way the media reports were transmitted. The presence of this spokesperson meant that journalists had a permanent source of information at all times. The spokesperson was also able to cater to the specific requirements of these journalists. This prevented the formation of alternative information circuits, as a result of which the authorities would have lost control over the accuracy of any news reports (as happened in 1993). Moreover, the spokesperson also took pressure off of the members of the provincial coordination committee, thereby allowing them to concentrate fully on running the actual emergency operations.

5 Disaster response

Coordinating emergency operations

The general structure of the relief operations — In response to the floods of 1993 and 1995, a wide-ranging relief operation was set in motion. For about ten days, several thousand relief workers were deployed throughout the country and all available means were mobilized to combat the inundations. The coordination of relief operations in response to the floods in 1993 and 1995 was carried out at the provincial policy level. For several days, disaster phase 3 was implemented in various provinces. From that moment on, authority lay in the hands of the provincial governor (see figure 3.4). A provincial coordination committee was created consisting of representatives from the various relief ser-

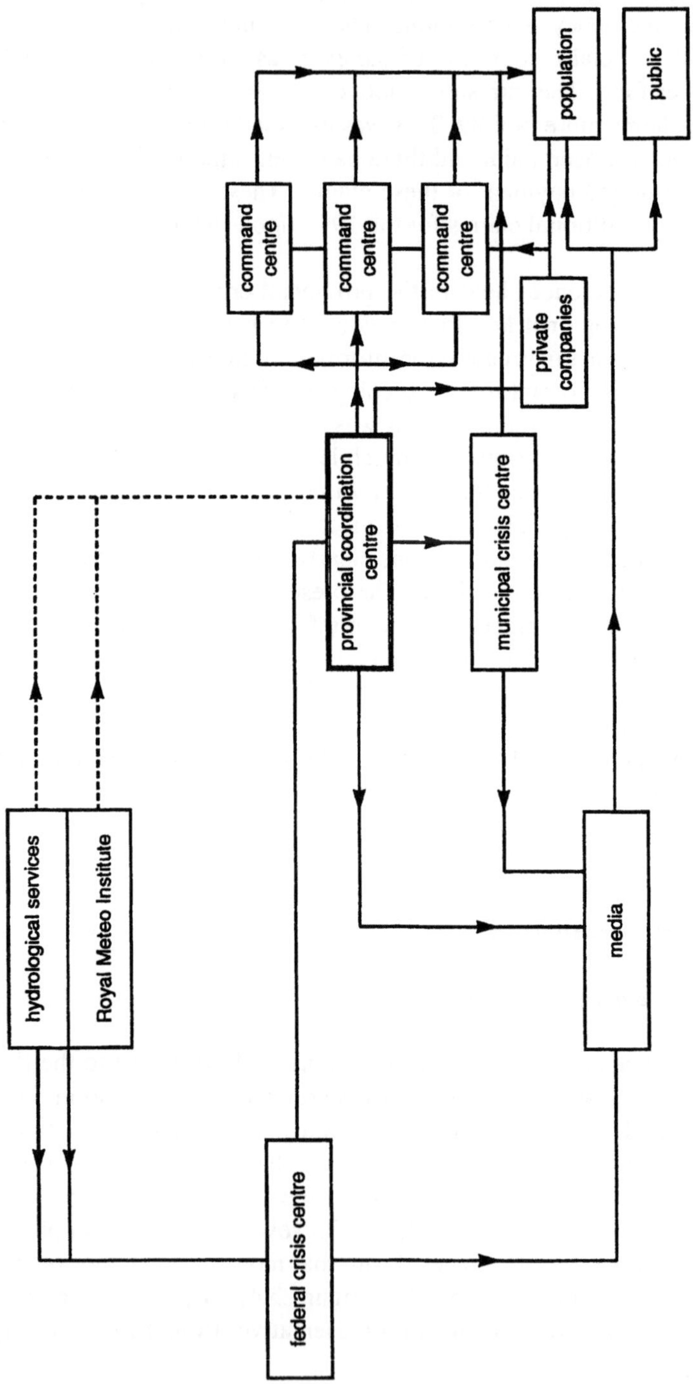

Figure 3.4 Emergency response during the 1993 and 1995 floods

vices, the municipal authorities, as well as a number of experts. Their task was to support the governor in evaluating the disaster situation, the deployment of equipment and personnel and the initiation of evacuation procedures.

Besides the provincial coordination committee, a command post was set up in the disaster area itself. Operational command rested with the fire chief, in consultation with the commanders of the relief services in the disaster area. The command post kept in constant contact with the coordination committee. Depending on the nature of the problems and the size of the disaster area, one or more command posts were set up in each of the provinces. In Limburg, it was decided to use one command post as a base for operations, whereas in Namur it was necessary to have several. Each command post was under the leadership of a fire chief and functioned under the authority of the provincial coordination committee.

Provincial coordination did not rule out coordination at a municipal or federal level. The municipal coordination committees were still able to function, but did so in accordance with the policy of the provincial coordination committee. They were above all active in the dissemination of information to the public and the coordination of operations of their own communal services. The federal crisis supported the provincial coordination committees by circulating information and delivering additional resources for rescue operations. A number of other organizations and services also played a role in the rescue operations: the hydrological services (for the measurement and forecasting of the water levels), the media (information to the affected population) and private individuals and organizations (distribution of food and equipment).

Why provincial coordination? — Given the enormous extent of the disaster area (seven of the nine Belgian provinces were affected) and the deployment of all available materials and personnel, the question is worth posing as to whether it may not have been useful to switch to disaster phase 4. If this had been done, coordination would have been taken over by the Minister of the Interior in close cooperation with the Government Crisis and Coordination Centre. In accordance with Belgian disaster planning, disaster phase 4 is declared from the moment a disaster impacts areas outside the provincial borders, which was clearly the case in both 1993 and 1995. The federal crisis centre did ask the provincial governors whether it would be wise to invoke phase 4, they believed this to be unnecessary. Their argument was that federal coordination would not make the rescue operations go any faster. The federal level possessed no overall disaster response plan capable of handling such a situation (there was one only for nuclear disasters). Coordination at federal level would merely burden and slow down the response. Moreover, provincial coordination was particularly suitable since each of the affected provinces was confronted with specific problem situations. The geographical spread of the floods was, moreover, such that demarcation of the disaster zones following the provincial borders made sense for topographical reasons. All of the required material and personnel were available at the provincial level and all of the affected relief services were briefed within the provincial coordination committee.

The coordination process in 1993

The initiation of disaster phase 3 — The initiation of provincial coordination in 1993 came neither from the municipal nor from provincial level. On 20 December 1993 the situation escalated so quickly on the Walloon water courses that the federal government called for provincial level coordination. As the geographical extent of the high water level spread throughout the afternoon, one by one the Walloon provinces decided to announce disaster phase 3. Around 2.00 in the province of Namur, the provincial coordination committee which would lead the ensuing assistance operations was set up with great haste. On 21 December, phase 3 was also announced in the provinces of Limburg and East Flanders.

At 12.30 on 21 December, the first crisis meeting of the Limburg provincial coordination committee began under the leadership of the provincial governor. It was the first time since disaster planning began that a relief operation of such magnitude and speed had been necessary. Although this scenario had been foreseen and prepared for in disaster planning, the actual implementation and further development of the relief operation was, for most of the people involved, a step into the unknown. In Limburg the warning of relief services went according to plan. The relief services reacted quickly and were rapidly deployed. In Namur, the initiation of the alarm procedure went less smoothly. Quick action was foreseen in the provincial disaster plan, but in practice the people involved lacked the experience necessary to react quickly and in an appropriate manner. The provincial policy-makers discovered that there was a substantial difference between procedures set out on paper and their actual implementation in a crisis.

Rescue operations — The magnitude of the emergency assistance operation initiated in 1993 was unprecedented. Fire service, civil defence, red cross, gendarmerie and local police personnel were engaged constantly in rescue operations. In a number of places, the army was also deployed. In Limburg, thousands of relief workers put in 12,610 full eight-hour working days. The army deployed more than 2,500 personnel. For days on end they worked to fortify and heighten the dikes and pump out cellars, houses and roads. In total, 190,000 sandbags were placed. Approximately 2,000 people were evacuated. The gendarmes and the police were busy with the influx of disaster spectators.

The cost of the relief operation in the Limburg Meuse was estimated at around BF 40,000,000. In the province of Namur, there were many cases of flooding of villages and town centres. Camp sites, meadows, roads, houses and companies were under water for days on end. A number of villages were completely isolated for several days. The deployment of the relief services was even more extensive than in Limburg.

In both provinces the provincial coordination centre was the central authority from which operations were controlled, making strategic decisions relating to evacuations, personnel and material deployment, sandbag placement, measures against disaster voyeurism, and food distribution to relief workers and the affected population. The existence of a provincial coordination committee offered a number of important advantages. First of all, it united officials from all of the affected relief services. It was possible to consider the measures to be taken and make sure that the intervention of the various relief services was coordinated. This created a synergy which offered an extra dimension when compared with the independent deployment of each relief service. By coming to agreement about their operations on a combined basis, overlap and misunderstanding on the ground could be avoided. Moreover, it was possible, via this provincial coordination, to deploy a far greater quantity of relief workers and materials. The committee was located at the fire service barracks in Maasmechelen, chosen for its central location. These barracks were centrally with the operational command post close by. This facilitated communication between the commanders and their respective relief services and between the relief services themselves. In the province of Namur, the provincial coordination committee met in the provincial buildings in the city of Namur. Given the large size of the disaster area, there would have been little point in locating the provincial crisis staff any closer to the disaster area. The command posts, however, were operative over the whole of the area of the province.

In both committees, key decisions were made by consensus. Unanimity was fostered by the time pressure inherent in the situation. It was also a matter of personal leadership by the provincial governors and their deputees, the district commissioners who act as chairmen of the coordination committees. In the province of Limburg, an important contribution was made by the engineers from the Albert Canal Service, whose technical expertise on the subject of dikes and water courses had a decisive impact on a number of important decisions concerning the deployment of personnel for the strengthening of dikes and the various evacuations.

Operational coordination rested with the fire chiefs. Also, the commanders of the various relief services were present, close to the action. In the province of Limburg, relief work was coordinated from a single command post while in the province of Namur, numerous posts were set up throughout the disaster area. After the provincial coordination committee had become operational, a number of crisis centres remained active in the municipalities. The initiatives and actions of the municipal and federal authorities were bound to rely on the policies of the provincial coordination committees. Among other things, they were responsible for passing out information to the affected population and for leading operations carried out by their own technical services. They played an important role in the evacuation of inhabitants. During disaster phase 3, the federal crisis centre went into "pre-alarm", monitoring the relief operations. During the entire period, the provincial coordination committees and the federal crisis centre exchanged information several times per day. In addition the federal crisis centre offered support to relief operations, delivering additional pumping equipment and sandbags, receiving the foreign press and distributing supplementary guidelines on how to handle disaster tourists.

The 1993 response: key problems

A relief operation of this magnitude was a new experience for those involved. Not surprisingly a number of problems and unforeseen circumstances arose during the course of this disaster.

Infrastructural problems — Both coordination centres in the province of Namur and the province of Limburg were insufficiently equipped. There were not enough suitable meeting rooms. In the province of Namur, rooms were cleared out and in the province of Limburg, the coordination committee had no separate meeting room. Moreover, officials had access to few or no maps or suitable supplies required to lead the coordination of the relief activities in cases of flooding. The secretariat of the coordination committee was improvized and undermanned, complicating the reporting process.

Communication problems — In Limburg the presence of a representative from Belgacom in the coordination committee ensured that all requirements concerning telephone communications were rapidly provided. Moreover, the various relief services were well-equipped with radio communications devices. They were, however, insufficiently informed about the possibilities of the new communications vehicle of the civil defence, which was underutilized. In Namur, personnel struggled with a shortage of communications equipment. Although the relief services did possess the necessary hardware, there was a dire shortage of telephone and radio links in the provincial coordination committee. Consequently, communications between the centre and field workers was fraught with problems in the initial phase. This led to a host of misunderstandings about the situation and the operations in the disaster area. Firstly, the relief services used separate frequencies for their radio links. In addition, the strength of the equipment used by some services as well as their superfluous drowned out the frequencies of others. Thirdly, in Namur, there was also a constant struggle with reception and transmission in lower-lying valley areas. A final communications problem was created by the insufficient communications skills of many relief workers. The volunteer workers in particular were often not well trained in radio communications, and proved unable to get the essential information across in a clear and meaningful manner. As a result information transmission between the trouble-spots and the central crisis staff became rather confused.

Coordination problems — Operational confusion shone through in the daily reporting of the relief services in the provincial coordination centre. The services did not always have a correct overview of the number of people and materials they employed in the field. These reporting problems were most apparent among those relief services which employed volunteers. The inadequate monitoring of materials and personnel led to misunderstandings. In Limburg, for example, a breach occurred in a dike when teams and material were standing idle elsewhere. In Namur some mayors made direct calls to the relief services to deploy material and personnel. This caused some misunderstandings within the provincial coordination committee which had not been informed of these interventions. Accordingly, the provincial coordination committee lost part of its grip on the operations in the disaster area. Manual sand-bagging proved overly time-consuming and food distribution to rescue workers and to the stricken population did not always run smoothly. It proved difficult to reach the various disaster areas because access roads were flooded. In addition, there was a lack of large-scale infrastructure needed to distribute the food in an efficient manner.

Evacuation problems — The decision of the provincial coordination centre to evacuate was broadcast repeatedly, but during its implementation, the authorities encountered notable opposition from the inhabitants. Some evacuations had to be carried out in difficult circumstances and under immense time pressure. On 22 December, for example, 2,000 people were evacuated in great haste in the middle of the night in Limburg, and in Namur a church wardens' retirement home also had to be evacuated during the night. These night-time evacuations were hard and risky for all concerned. Most of the evacuations were carried out in a highly tense atmosphere, since there had been no advance information on the possibility of evacuation and the measures to be taken.

During the first few hours of the evacuations a number of problems arose: the registration centre and the reception centres were overrun. Furthermore, the dispersement of the evacuees over different reception centres proved impossible to supervise. Some evacuees attempted to return. In their haste many of the evacuees forgot their identity papers and everyday requirements (medications, baby supplies, clothes). In the province of Limburg, nearly all of the evacuees who could not be put up in families or elsewhere were taken to one reception centre by the Red Cross after registration. In the province of Namur, evacuees were put up in various small reception centres spread over the province. These were predominantly holiday villages, hotels and schools.[15]

An important bottleneck was caused by the refusal of many to heed the order to evacuate. For the most part, forced evacuation was the only option remaining, but the provincial coordination committee believed forced evacuation too time-consuming and risky for the relief workers. Thus nobody was obliged to be evacuated. Of all those unwilling to leave, the most hesitant were those who would have to abandon their livestock. Those remaining behind were registered and were expressly warned that they did so at their own risk. Had the worst indeed developed, there would have been serious problems of responsibility and liability.[16]

Strategic versus operational command posts — From a practical point of view there was hardly any clearcut division of labor between the provincial coordination committee and the command post in Limburg. Provincial policy makers were besieged with problems which should have been solved at the command post, while executive relief workers became needlessly involved in the strategic decision making process. The provincial coordination committee regularly took over the role of operational command post, deliberating about issues like sandbag supplies. In the initial phase, this led to confusion and delays in the deployment of personnel and materials.

The cause of this problem was twofold. One, there was no physical distance between the two units. Both were concentrated in the fire station barracks in Maasmechelen. And two, there was also a pronounced absence of the required infrastructure. The provincial coordination committee could not obtain the use of a separate meeting room, nor did the various relief services have their own command posts with the required personnel, equipment and means of communication. Accordingly, there were simply too many people present in and around the coordination committee which contributed to add to the sometimes chaotic condition of its performance.

Disaster tourists — The intensive media coverage encouraged the presence of a host of voyeurs and inquisitive individuals in the disaster areas which greatly hampered relief operations. In Limburg, the stream of curious onlookers constantly blocked the access routes to transport of equipment and relief workers. The Gendarmerie and local police had to set up additional roadblocks and mount controls. Moreover, many of these onlookers put themselves in danger by venturing onto saturated dikes. In a number of places in Namur there were physical confrontations with these onlookers.

Conflict among relief workers — In general cooperation between the various relief services went relatively smoothly. The commitment of many volunteers proved an essential contribution to the successful execution of relief operations. Nevertheless, there was friction between civil defence and fire service relief workers. In particular, in the province of Namur the position of the civil defence service was strongly criticized and certain fire service units refused to cooperate further with them. The explanation for these tensions can be found in the partially overlapping tasks of the two relief services as well as in their differing method.

Conflict among authorities — Aside from problems associated with some municipal crisis centres not being integrated from the outset in the provincial coordination structure, the most serious conflicts between authorities centered on the evacuations. In both Namur and Limburg, the provincial coordination committee gave the order to evacuate several times, but mayors delayed or blocked the execution of the order. The final authority to evacuate inhabitants is the mayors'. Without their approval, the provincial coordination committee could not move to evacuate. In spite of the threatening Meuse water, certain mayors hesitated to evacuate, mostly, because many of their townspeople refused categorically to be evacuated and put pressure on their mayor to be allowed to stay in their houses (often with their cattle). The mayors' local presence thus tied their hands when it came to taking decisive preventative action.

Official visitors — The disaster areas were regularly visited by dignitaries and politicians wanting to ascertain the seriousness of the situation. There were official visits from the King and Queen, Prince Philip and a number of federal, communal and regional officials. There is no doubt that these visits were important morale boosters during a time of trouble. Nevertheless, they demanded a good deal of attention, personnel and equipment at a time when the relief services are already overworked and overloaded. They also involved additional work for the provincial coordination committee. The preparation and the accompanying of each visit took up several hours, at a time when the pace of events was hectic and manpower short.

The coordination process in 1995

The initiation of disaster phase 3 — In 1995 the initiation of disaster relief was prompt and controlled. There was no sign of the nervous atmosphere which reigned during the initial hours in 1993. In 1995 the final decision to begin the provincial coordination lay in the hands of the relevant governors. This was made possible by a timely advance warning of the approaching water. The 1993 experience of a quick escalation of the situation had led to a greater degree of awareness at all levels of government. The situation was in hand hours, even days, in advance. In Limburg there was a preparatory meeting of the provincial crisis committee on 23 January, and the required preventative measures were taken. In Namur the governor called the affected relief services together, a few days before the announcement of disaster phase 3, to discuss the situation. On 26 January, disaster phase 3 was announced by the governors. The relief services reacted appropriately. In Limburg the provincial coordination committee met for the first time at 9.30. During this meeting the most important measures were taken for the initiation of the relief operation. In Namur, the start-up of provincial coordination went quickly and smoothly. The launching of relief operations went according to plan. Disaster phase 3 was operative from 26 January until 3 February 1995.

Rescue operations — The extent and intensity of the 1993 and 1995 floods were comparable. The problems were concentrated, for the most part, in the Meuse basin. The provinces of Namur and Liège and to a lesser extent the province of Limburg were worst hit. Again, large-scale relief operations were in order. For about ten days, the relief workers of the fire service, gendarmes and police, the Red Cross and the civil defence service were fully engaged. Similarly, the technical services from the communes, the Public Centres for Social Welfare, engineers and personnel from various waterways and the telephone and electricity companies played vital support roles.

In 1995 the general coordination of the relief operations was handled at the provincial policy level, i.e. by the provincial governor in cooperation with the provincial coordination committee. Just as in 1993, the Limburg provincial coordination committee met in the vicinity of the disaster area at the fire service barracks in Maasmechelen. In Namur, the provincial coordination committee met again in the provincial buildings in the city of Namur, but this time in the rooms of the permanent crisis centre. In Limburg, relief work was predominantly geared to strengthening and heightening the dikes, placing sandbags, pumping out superfluous water, carrying out a few evacuations and suppressing disaster tourism. From the third day, the operations proceeded routinely. In Namur, the provincial coordination centre met once a day at a fixed time. The relief work was primarily aimed at pumping water out of houses, company premises and streets, placing sandbags, carrying out evacuations, providing meals and drinkable water and curbing disaster tourism.

Comparing the 1993 and 1995 experiences — In general, relief operations in 1995 ran smoothly, expertly and on time. No doubt, all concerned benefitted from the experience gained during the highly similar relief operations in 1993. In addition and in contrast to 1993 the relief services were warned on time, and were thus able to prepare themselves much better. Without doubt this was partly due to learning processes that had occurred following the 1993 disaster. We review some of the main differences between the two response episodes.

With regard to infrastructure, important efforts were made in both provinces. In Limburg, a purpose-built meeting room was set up for the coordination committee containing area maps and other provisions. Also, relief services were given their own rooms in the vicinity of the fire service barracks. In Namur a more radical improvement was made in 1994, a permanent crisis centre was installed. This centre consists of several meeting rooms for the different relief services and is equipped with the necessary communications devices, maps and other requirements. This crisis centre is designed to allow for extended crisis operations.

The evacuations ran smoothly. Relief services and the population were prepared for the evacuation this time. Reception and registration of the evacuees were well organized. Already during the pre-alarm phase, information leaflets were handed out to populations in the province of Limburg, and the Red Cross set up four registration centres and possible reception centres.

This time, disaster tourism was suppressed efficiently. After the 1993 floods, the governor of the province of Limburg had urged the Minister for Internal Affairs to take measures in order to alleviate disaster tourism. Guidelines were established for local police regulations. During the 1995 floods, disaster areas were swiftly declared off-limits for anybody not directly involved in the relief operation. Any breach of the regulations was punishable by fines. The gendarmes and the police set up roadblocks sealing off the disaster area.

The coordination of relief operations proceeded more efficiently and in a more controlled fashion than in 1993. In both provinces, the coordination committee achieved a more comprehensive overview of operations in the disaster area. This was made possible by better communications between the field and the provincial crisis staff. Access to the centres was more carefully guarded and separate, quiet meeting rooms were available for the policy makers. In Namur, especially post-1993 investments in communications equipment paid off. The daily briefings of the heads of the relief services in the provincial coordination committee gained importance. The centres were thus proveded detailed and accurate overviews of the deployment of personnel and equipment. In Limburg, much work had been done following the 1993 floods on setting up a communications plan. In it, the various relief services were allocated specific frequencies for their radio communications. The plan also outlined their interorganizational communications, as well as the information exchanges between the relief services and the provincial crisis staff.

Three watch services were set up for dike surveillance on the Limburg Meuse. Three fire service jeeps, each manned by a fireman and a dike watchman from the Meuse and Albert Canal Service, patrolled their sections for eight days. They were in constant radio contact with the fire service barracks of Maasmechelen and Maaseik. This was all part of an effort to avoid the false reporting of dike breaches, which had occurred frequently in 1993.

In order to promote cooperation among the relief services, the provincial coordination committee of the province of Namur divided the disaster area into operational zones. The relief services, specifically the fire service and the civil defence service, were each allocated a certain territory. In this way, overlap and subsequent conflict among the relief services could be avoided.

The 1995 response: key problems

A recurrent bottleneck was the refusal of a large number of those involved to heed the order to evacuate. In 1995, as in 1993, may people did not move until they were forced to. Those remaining behind were registered and expressly informed that they stayed behind at their own responsibility. In 1995, there were conflicts among authorities resulting from the evacuations. In the province of Namur, evacuations were repeatedly halted or delayed by the mayor responsible even though the provincial crisis staff had given the order. The mayors were strongly influenced by the strong local opposition to being evacuated. In Namur, a problem of authority and coordination existed with respect to the deployment of relief workers and equip-ment during operations. Some mayors directly addressed those responsible for the relief services with requests for pumps, boats or sandbags. Finally, numerous official visits again put a strain on the policy makers and the relief services at a time when they were already overworked.

6 Recovery and damage compensation

Assessing the damage sustained as a result of the 1993 and 1995 floods is a difficult and delicate task. Certain types of damage proved difficult to measure or were not precisely quantified since they were not relevant to a damage claim. The figures quoted below should therefore be interpreted with some caution.

In Limburg, direct material damage to private and public property came to more than BF 100 million following the 1993 flood. Damage to agricultural property (crops, livestock, pastures and agricultural equipment) came to around BF 18 million. The overall material damage sustained (including commercial losses, the cost of clean-up operations, longer term damage, etc.) was estimated at around BF 300 million. In addition to the material damage suffered, those affected also sustained other forms of loss: one person and hundreds of animals were drowned, and some 2,000 people and several hundred animals had to be evacuated. Victims of the flooding also suffered psychological stress, fear and uncertainty about what would happen to them, upheaval and the disruption of their daily lives and of preparations for the festive season. The cost of repairing the infrastructure damaged by the 1993 floods (dikes, locks and flood barriers) throughout the whole of the Flemish region was set at approximately BF 1.5 billion. This allone would absorb the entire budget set aside by the Flemish government for water management in 1994.

In Limburg, it was estimated that direct material damage following the 1995 floods totalled some BF 50 million. Damage to agricultural property came to around BF 150 million. In comparison to the 1993 floods, only a few dozen families were evacuated this time. Although the intensity of the disaster was at least as severe as in 1993, there was far less damage in 1995 than in 1993. This can be explained partly by the strengthening operations which had been carried out on the dikes, the increased pumping capacity and the preventive measures which were carried out in good time by the population.

The National Disaster Compensation Fund

The National Disaster Compensation Fund was established by law to provide financial compensation to victims of natural disasters. Disasters are defined under this law as 'natural events of an exceptional nature or of a severity that cannot be foreseen, and which have caused major damage, notably earthquakes or tremors, flooding or other high-water having a calamitous effect, hurricanes or other high-force winds.' The law does not apply in cases where the losses are caused by risks which can ordinarily be covered by insurance contracts, such as fire, lightning or hail. Before the National Disaster Compensation Fund can be activated, the incident in question must be officially recognized as a disaster. This recognition is given by the Minister of the Interior.[17]

The procedure for activating the Fund is as follows. As soon as an incident occurs which can be designated a disaster, the mayor of the affected munici pality immediately gathers together as much information as possible. All of the affected inhabitants in the municipality are asked to make a general inventory of their damaged property, but this does not automatically imply a recognition of an individual's right to claim compensation. A summary is then drawn up of the damage sustained in each municipality. This information is submitted to the provincial governor who makes a detailed report for the Minister of the

Interior.[18] The Minister then decides whether to submit the matter to the Council of Ministers in order to have a national disaster declared.

As soon as the official disaster decree has been published in the Belgian Official Gazette, victims have six months in which to submit their claims for private property. Only physical damage to property will be compensated, not damage to morale, financial loss, commercial operating losses or physical injury to persons. The application must be sent to the provincial governor on specially provided forms together with the necessary supporting documents. The provincial governor appoints an expert who will determine the final costs of the damage, and the amount of compensation to be paid out by the state is calculated on the basis of this amount. The governor decides on the amount of compensation to be paid out. The state will never repay all the losses, but will help only to meet the costs of repair.[19] This payment consists of an amount paid out in compensation for repairs and of a loan for repairs. In cases of pressing need, the provincial governor can also allow pre-payments to be made.

Any victim awarded financial compensation is obliged to spend the compensation and the loan on the repair of the damaged property within three years from the date of the governor's decision. No compensation will be paid without proof of reinvestment. The governor's decision will in turn be subject to an administrative, legal and technical assessment by the Federal Ministry of Traffic and Infrastructure which is responsible for managing the Disaster Fund. Once a form has been fully processed, the payment orders are executed.

The state can also award ad-hoc subsidies to the regions, communities, provinces, municipalities, inter-urban districts, OCMWs (Social Welfare Offices), church commissioners and polder and waterway management boards to help repair the damage to public property which they are responsible for managing. Applications for subsidies of this kind must be submitted directly to the Disaster Fund Office at the Federal Ministry of Transportation and Infrastructure.

National compensation in 1993

During the initial days of the 1993 floods, questions were already being asked about who would compensate the damage sustained.[20] During a visit to the affected areas, a cabinet minister suggested that the situation would quite likely be designated a disaster and that victims could apply to the National Disaster Compensation Fund. This was officially confirmed at the end of March 1994. From this moment on, victims of the disaster were allowed to submit compensation forms.

A total of 3,632 compensation forms were submitted to the National Disaster Compensation Fund. The worst affected area was the province of Namur, which yielded 1,537 claims, followed by the province of Liège with 1,005 claims.

Several hundred claims were submitted from three other provinces, as well as small numbers from four other provinces. At the end of October 1995, a total of 1,899 binding compensation decisions had been taken, involving a total amount of BF 252 million. There were differences between provinces in the processing of the claims. In the Flemish provinces, the processing of the claim forms was almost complete by late 1995. In Wallonia, which suffered more damage, the procedure has been slower. In the provinces of Namur and Liège, the authorities have been awarding many more pre-payments. The policy of the governors in these provinces was to provide financial aid to as many people as possible at the earliest possible opportunity. In Namur, no less than 1,279 claims have been awarded a pre-payment, involving a total of more than BF 76 million. However, by the end of 1995, only 84 definitive decisions had been made, on a total of 1,537 claims. Moreover, the massive use of pre-payments imposed an additional burden on the provincial administration, since the same forms were being processed several times.

Twenty-nine forms were submitted for compensation for the repair of public property involving a total amount of almost BF 200 million. Most of these were from municipal authorities. Some BF 26 million was awarded in compensation, BF 8.5 million of which had actually been paid out by the end of October 1995.

National compensation in 1995

During the initial days of the 1995 floods, the government again declared that the affected areas would be able to make a claim under the National Disaster Compensation Fund. It also announced measures to speed up the processing of claims. This time, 2,400 claims were submitted. Once again, the province of Namur was the worst affected (1,510 claims), followed by the province of Liège (441 claims). Limburg came in fourth with 126 claims. At the end of October 1995, there were 361 definitive compensation decisions involving a total amount of some BF 44 million. By then, 228 payment orders had been transferred to the National Disaster Fund, and BF 13 million had actually been paid out in compensation. About BF 3 million pre-payments had been made. Once again, there were differences between the various provinces. In the Flemish provinces, around 75 percent of claims were processed. In Wallonia, which was more seriously affected in the 1995 floods than Flanders, the process is proceeding very slowly. The need to process claims from 1995 on top of those from 1993 has severely overburdened the provincial administration. Major delays have been the result. In the province of Hainault, no definitive decisions have been made, either about final payments or about pre-payments. In the province of Liège, only 97 decisions have been made regarding more than 400 claims, and in Luxembourg, only 23 definitive decisions have been made regarding 156 claims. The situation in the province of Namur is equally serious. By the end of October 1995, only 3 definitive decisions had been made involving a total of BF 400,000 in damage compensation. Yet again, many pre-payment decisions have been made, and this has further burdened the administrative processing of claims. This time, 19 claims were submitted for compensation for repairs to public property involving a total of some BF 60 million. Most of these claims were submitted by the municipal authorities.

National compensation: A slow process

The processing of the claims and the payment of compensation to victims has been progressing slowly and with difficulty for several years. Those affected often do not receive compensation until many years after the disaster. This is due to a variety of reasons:

- The government does not have enough specialized administrative personnel. The entire claims process imposes a severe burden on the administrations of the various levels of government concerned.
- Before a claim can be fully dealt with, it must be processed by many different services at various administrative levels. The municipal administrations, for example, must provide people with information and assistance in filling out the forms, and they do not always have the necessary staff to do so. The provincial authorities are assisted in processing the claims by additional staff temporarily recruited by the federal government. However, the practice of recruiting additional personnel to deal with each disaster is a time-consuming business and leads to delays. At the federal level, three departments are involved in processing damage claims: the Ministry of the Interior (which is responsible for recognizing the incident as a disaster and makes available administrative staff to the provincial governors), the National Disaster Compensation Fund (which is administered by the Ministry of Transportation and Infrastructure) and the National Disaster Fund (which is administered by the Ministry of Finance).
- Once an event has been recognized as a disaster, victims have six months in which to submit their claims. Experience has shown that most do not submit their claims until the last moment.[21]
- The forms supplied for making a claim for damages are extremely user-unfriendly. Even with the explanatory notes provided, they are still very difficult and time-consuming to complete. Consequently, submitted forms are rarely filled in correctly or completely.
- The law does not specify any maximum period within which the provincial governor must make his decision.

- Once the provincial services have completed their investigation, the claim and the provincial governor's decision are assessed by the Disaster Fund Office. This Office regularly asks for an adjustment or an annulment of the claim, and sometimes even submits an appeal to the Court of Appeals.
- Once a final decision has been made, the victim has three years to submit his proof of reinvestment. He will be paid only when this has been submitted.
- The National Disaster Compensation Fund has no resources of its own, so any payments are wholly ascribed to the budget of the Ministry of Finance. Over the last 20 years, the Disaster Fund has repeatedly imposed a heavy burden on the already debt-laden Belgian economy.

Although the processing of the claims following the 1993 floods was handled more efficiently than it had been in the past, the federal government is still looking for ways of further speeding up this time-consuming procedure. During the 1995 floods, the public was assured everything would be done to make that compensation procedures be speeded up. A "Task Force" was established to monitor the processing of claims and to promote a more efficient operation of the National Disaster Compensation Fund. For example, it decided to reduce the period for submitting a claim in 1995 to three months instead of the six. It also issued a guide aimed to provide claimants with clearer information. The federal government also took the necessary measures to speed up the completion of claims at the provincial level. In each province, teams of experts were made available, initially to assist the municipalities in providing information and assistance to the affected populations. In the most severely affected municipalities, information days were organized and a municipal information help desk was set up. During a second phase, these experts were involved in helping to process forms at the provincial level.

Additional compensation

The Emergency Aid Fund (Flemish region) — After the 1993 floods, the government of the Flemish Community established an emergency aid fund.[22] This fund provides a supplementary subsidy to Flemish municipalities affected by exceptional and unforeseen circumstances with severe repercussions for the municipal budget. Applications for supplementary funding by municipalities must be made within four months and must be fully substantiated. The decision about whether to award a subsidy is made by the minister following the advice of an expert committee. It was decided not to opt for the time-consuming submission of invoices but instead to work on a fixed payment basis. The contributions to the fund must be paid by the Flemish municipalities in accordance with the principle of inter-municipal solidarity. The fund was activated in 1995.

Special pre-payments (Wallonian region) — The time-consuming procedure involved in the National Disaster Compensation Fund prompted the Wallonian Regional authorities to introduce a special scheme under the order governing the budget for expenditure. Under this scheme, the Regional Executive can award a guarantee and can assume the interest payments on the loans made to municipalities affected by a disaster. In this way, victims can be awarded pre-payments on the amount which will later be paid out by the National Disaster Compensation Fund. In addition, since a significant number of camping sites, hotels and catering establishments were affected by the flooding, the Wallonian authorities awarded special subsidies to this sector.

The European Community — Shortly after the 1993 floods, the European Union stated that it would be making available ECU 500,000 for the victims. This aid was distributed by the Belgian Red Cross. In doing so, an attempt was made to compensate damage not covered by the National Disaster Compensation Fund.

Towards private insurance?

In 1992, in view of the ongoing organizational and financial problems associated with the National Disaster Compensation Fund, the Belgian government decided to gradually privatize the Fund. The Committee for the Insurance Sector drew up a proposal to expand the coverage provided by fire insurance, first to include storm damage, then earthquakes and finally also flood damage. On 14 January 1994, the Council of Ministers approved the proposal to make it obligatory for fire insurance to cover storm damage. The situation with regard to earthquakes and flood damage is more complex. According to a recommendation by the Committee, the government should include these under fire insurance. The insurers could then use the premiums collected to build up reserve funds, which could be used to expand the coverage proportionally.

The insurance companies are reluctant to accept these proposals. According to them, using fire insurance to cover the risks of flooding is technically almost impossible to do at present since flooding is a fairly high-probability risk which is difficult to spread. As long as fire insurance in Belgium is not obligatory and as long as customers are free to construct their own policies, there is no point in using such policies to cover the risk of flooding. Those living in flood-free zones will naturally not subscribe, thereby making the premiums for those in flood risk areas unaffordable. A further problem is that Belgian insurance companies transfer their risks to foreign insurers. This is relatively straightforward where glass and fire insurance are concerned because these risks can be calculated. The risk of flooding, on the other hand, is unpredictable and therefore practically impossible to reinsure.

The proposals which insurers have put forward so far with regard to the inclusion of the risk of earthquakes and flooding under fire insurance have generated considerable debate. The government has asked the insurers to formulate a new proposal based on the French system. In France, an Act of 13 July 1982 introduced a system of private insurance for both damage to property and for commercial operating losses. This insurance is underwritten by the state.

Notes

1 Througout this chapter we shall use the term Meuse and not Maas, since this correponds with the English spelling of the river's name.

2 For this study, all possible information was collected from the parties involved in the emergency operations: the policy executives and their administrations (the municipalities and Social Welfare Offices, the provincial administrations of Limburg and Namur, the services of the Flemish and Wallonian Regions and Communities and the services of the federal government), the emergency aid organizations (civil defence force, army, fire service, state police force and Red Cross) and the various media representatives (television, radio and newspapers). Initially, this was done by contacting those involved and asking for information about their activities following the emergency operations. In a second phase, the various activity reports, minutes of meetings, updates, expense statements, evaluation reports and other documents were analysed. The news bulletins on the floods transmitted by the various media were also analysed. In a third phase, an extensive, supplementary questionnaire was sent to the heads of the emergency service organizations who had been active in the province of Namur. This questionnaire asked specific questions about their involvement in the emergency aid operations, their experiences and the decisions they had made. During a fourth phase, the study was supplemented by some 20 interviews with key figures who had played a role during the emergency operations, either in terms of the implementation of policy or as part of an emergency service. During the study, we were given the opportunity to study documents at the Government Coordination and Crisis Centre, and we were also assisted by the civil defence services at the Ministry of the Interior.

3 Its main tasks include collecting information about crisis situations, analysing this information and making it available to policy executives; making available the necessary infrastructure and resources to the federal government to enable it to manage a national crisis; taking necessary measures as outlined in the disaster plans; and ensuring the provision of clear, distinct information.

4 *Ministeriële Omzendbrief van 11 juli 1990, artikel 14*, (Ministerial Circular of 11 July 1990, Article 14).

5 The "Seveso directive", issued by the Council of the European Union on 24 June 1982, is concerned with the subject of risks of severe accidents involving certain types of industries. The directive dictates that certain industrial compagnies take preventive measures to minimize the risk of disaster as far as possible (so-called 'on site safety precautions'). Moreover, the directive dictates both compagnies and local authorities to undertake several planning and public relations activities (so-called 'off site safety precautions').

6 *Koninklijk Besluit van 22 april 1988 tot vaststelling van het type, de modaliteiten en de procedure van de informatie die moet verstrekt worden wanneer een zwaar ongeval zich voordoet bij bepaalde industriële activiteiten (Belgisch Staatsblad 03.05.1988)*, (Royal Decree of 22 April 1988 to Establish the Type, the Modality and the Procedure of the Information that must be Provided when a Serious Accident Occurs Involving Certain Industrial Activities).

7 Provinciebestuur Limburg, *Algemeen Noodplan*, Hasselt, 1993, 2 volumes.

8 Province de Namur, *Plan General de Secours de la Province de Namur*, Namur 1988.

9 During this period, there was flooding in several locations. One or two camping sites were flooded on the banks of the Semois and Lesse. At Dinant, the Meuse flooded its banks in several places. These local floods did not give cause for concern, since they regularly occur at times of high water, and there are many camping sites located within the winter flood plain of these rivers. The sites are regularly inundated during high water.

10 BRTN radio news bulletins of 19 December 1993 (20.00, 23.00 and 24.00). RTBF radio news bulletins of 19 December 1993 (20.00 and 22.30).

11 A water level of 45.5 TAW at Smeermaas prompts an "early warning" among the emergency services and a level of 46 TAW or higher prompts an "emergency alert" situation.

12 On 20 December 1993, around 13.00, the Diho received an initial fax from the Dutch Department of Public Works. From the early warning phase onwards (13 December 1993), the Sethy regularly sent bulletins to the Dutch Department of Public Works. For the Maas and its tributaries, an early warning was sounded on 19 December 1993 and a general alarm around midday on 20 December 1993.

13 In Belgium, water levels are measured using the TAW standard. TAW stands for Tweede Algemeen Waterstand [= Second General Level]. The highest measurement recorded for the Maas (in December 1994) was 47.32 TAW. This corresponds to 44.65 NAP (New Amsterdam Level), the measurement standard used in the Netherlands.

14 For example, one radio bulletin incorrectly reported that the town of Elen had been evacuated, and a newspaper report stated that the coordination committee was considering blowing up the dikes at Uikhoven to relieve the pressure on the dike around Maaseik, and so spare the town of Maaseik by "sacrificing" the smaller town of Uikhoven. According to a report in 'De Morgen' newspaper, contaminated soil from an old zinc factory in Dilsen-Stokkem was used to reinforce the dikes.

15 The provincial coordination committee was surprised to learn when it contacted the holiday village in Hastière that the entire centre had already been fully booked by the mayor of a neighbouring French commune with a view to evacuating the inhabitants of his own village.

16 The law does provide for enforced evacuation, but there are apparently no examples in (Belgian) jurisprudence which provide more clarity on this issue. Further study is therefore required.

17 This is done by means of a Royal Decree specifying the date and time of the disaster and defining the extent of the affected area. The 1993 floods were, for example, covered by a Royal Decree of 1 March 1994 (Statute Book 23.03.1994) while the 1995 floods were covered by a Royal Decree of 24 February 1995 (Statute Book 03.03.1995).

18 This report must indicate how many persons were affected by the disaster and the nature and scope of the damage sustained. It must also give an estimate (in figures) of the damage and include a list containing the names of all the institutions which took part in the relief operation.

19 Compensation to meet the cost of repairs is calculated by means of coefficients on the basis of the net losses sustained by the claimant. The first risk element is BF 10,000 and any net losses in excess of BF 10,000,000 will not qualify for compensation. The maximum compensation paid is BF 2,492,000.

20 This information concerning the processing of the damage compensation forms after the 1993 and 1995 floods was made available to us by the Disaster Damage Office at the Federal Ministry of Transport and Infrastructure and the Disaster Fund Office at the Ministry of Finance. The figures quoted relate to the situation recorded on 31 October 1995.

21 Following the damage caused by the 1990 storm, over a quarter of all claims submitted in Limburg were received on the final day of the application period.

22 Royal Decree of 6 July 1994 governing various provisions for the financing of the municipalities (Statute Book 27.07.1994).

Chapter 4

FLOOD MANAGEMENT IN THE NETHERLANDS

Uriel Rosenthal, Menno van Duin, Michel Bezuyen, Marie-Louise de Vreeze-Verhoef and Paul 't Hart

1 Introduction

This chapter provides an analysis of the 1993 and 1995 riverine floods in the southern provinces of the Netherlands. In 1993, flooding occurred mainly in the river Meuse and remained confined largely to the southernmost province of Limburg, whereas the 1995 floods occurred in all of the main rivers in the Netherlands and affected six provinces.

The occurrence of two flooding episodes in a relatively short amount of time and in more or less the same region enables a two-fold comparative study design. On the one hand, we will be able to compare disaster management in the province of Limburg through time (e.g. using 1993 and 1995 as measurement points), and also compare simultaneous disaster management processes in two different provinces (e.g. Limburg and the province of Gelderland in 1995).

Although we have taken the two provinces as our main units of analysis, the units of measurement are the different flood management response regions within the provinces. This also allows us to observe similarities and differences within a province. Consequently, data has been gathered systematically for each of six regions: Southern, Central and Northern Limburg in 1993; and Nijmegen, Rivierenland and Arnhem for 1995. The degree of variety in disaster planning and response which can be found between those regions is presumably limited because they are embedded in the provincial and national system of government. The report will make it clear that, nevertheless, there were noticable inter-regional differences in flood preparedness and response.[1]

2 The 1993 and 1995 riverine floods: an overview of events

The 1993 floods

Indications of threat — In December 1993, just as in every other year, people in the province of Limburg were preparing for Christmas. Meanwhile, heavy rainfalls in France and the Ardennes in Belgium caused the water level of the Meuse to rise to a considerable height. On 14 December, the department of Public Works and Water Management (RWS), a section of the Ministry of Transport and Public Works, sent out an early warning signal to the regional fire brigade about the rising water levels. No one really worried, because, after all, the rise of the water level was a regular event in this region. The same situation occurred in 1980, 1984, 1986 and the beginning of 1993 and caused no serious problems. This reserved attitude seemed justified when RWS withdrew the early warning on 16 December. Nevertheless, the real problems were only to begin shortly after the withdrawal.

The water level of the river Meuse started to rise again on Saturday 18 December. The rising water was caused by continuous rainfall in the northern part of France and the Belgian Ardennes. The Belgian news bulletins showed pictures of floating caravans, following the overflowing of the rivers Sambre and Ourthe. That same night (Sunday 19 December 1995) several other alarming signs were given by Belgian correspondents.

Threat escalation — On Monday 20 December it became clear that the water level of the Meuse was reaching a threatening height. At 20.00, a crisis meeting was arranged in Itteren and Borgharen, two small villages at the Dutch/Belgian border.[2] Two hours later the so-called 'crisis level' was reached at the dam of Borgharen. Emergency services were alarmed in several parts of the province of Limburg. Shortly after the alarm had been given, several parts of the villages of Borgharen and Itteren were slowly invaded by the incoming water. During the course of Tuesday other villages along the Meuse were flooded. The situation deteriorated further on Wednesday 22 December.

The water level of the river was higher than expected, and the amount of people that had to be evacuated rose to 3,000. Along with the operational services, the military service also got involved in the rescue operation. Meanwhile, more villages suffered from the rising water in the river. Nevertheless, many people in Limburg still underestimated the threat of the flood and postponed the moment of evacuation as long as possible.

The water left a trail of destruction. Infrastructural utilities started to fail, farmers saw their cattle drown and businesses feared for bankruptcy. This trail of disruption immediately initiated a discussion on who would compensate the damages even though emergency services were still fighting against the water. Insurers declared that natural disasters were excluded from insurance coverage. The central government in The Hague, concerned with the faith of the people in Limburg, publicly considered the possibility of using the National Disaster Fund for damage compensation.

Other parts of the province of Limburg became affected by the floods. The mayor of Roermond declared the region of Central Limburg a disaster area on Thursday 23 December. Citizens were evacuated by helicopters, boats and military vehicles. Public authorities constantly emphasized the severity of the situation as it grew worse by the hour. The Queen and ministers of Public Works and Agriculture visited the disaster area. The figure of evacuated people had reached 10,000 and thousands of hectare land had been flooded. The minister of Transport and Public Works and the governor of the province of Limburg formed a special commission to study the future possibilities of controlling the Meuse.

A national disaster — The hard-hit area was officially declared a national disaster on Christmas day. This meant that damages would be compensated for as long as the budget would allow this. That same evening, roughly a fifth of Limburg was inundated. During the following days many dignitaries visited the disaster area to form their own opinion about the situation. Meanwhile, citizens were getting annoyed with the so-called 'disaster tourists'. Mayors proclaimed a state of emergency to prevent looting and to give the police a broader mandate for dealing with disaster tourists.

The minister of Interior visited Limburg on Monday 27 December. She did not make any promises, but declared that 'the government would "not be childish" in compensating damages.'[3] On Tuesday 28 December, the water level of the Meuse finally started to decline. The province began with the inventorization of the damage. Registration forms could be collected at the province. From Wednesday 29 December on, the various emergency services started reducing staff.

Another flood? — Soon after the first floods, another warning was issued by RWS on 30 December. The water- level of the Meuse was rising again due to heavy rainfalls in the northern part of France and the Ardennes. A second flood seemed within a matter of time in down-stream Limburg. Measures were being taken everywhere and emergency services once again mobilized their staff. This time it turned out to be false alarm.

After the risk of a second series of floods had abided, public discussion focussed on the compensation of damages. A national telecast was set up to collect money for the Disaster Fund. Estimations of the amount of damage vary from DFL 150 to 200 million (approximately 95-120 U$). A special meeting was arranged in the Parliament's Second Chamber to discuss the arrangements for damage compensation.

Characteristics of the 1993 floods — The riverine floods of 1993 had a gradual onset. It took the water four days to flow from the most southern to the most northern village in the province. Changes in the water level gave warning and allowed for preparations. Table 4.1 presents the different times and dates at which the highest water level was reached in sixteen municipalities along the river Meuse. In Figure 4.1 these municipalities are marked on the map of the river basin Meuse. The gradual onset of the riverine flood, as will be indicated in our analysis, had some implications for the response to the disaster, notably communication and administrative organization.

Table 4.1 Highest water level of the river Meuse in 1993

Location	Passage highest water level Date	Time	Highest water level (+ NAP)
Borgharen-Dorp	22-12	8:00	45.90
Elsloo	22-12	9:00	40.50
Grevenbicht	22-12	16:00	32.80
Stevensweert	22-12	18:00	25.36
Heel-boven	22-12	21:00	22.81
Roermond-Buggenum	23-12	19:00	20.52
Belfeld	23-12	21:00	19.05
Venlo	23-12	21:00	18.35
Well	24-12	18:00	15.34
Sambeek	25-12	4:00	13.92
Grave	25-12	17:00	10.39
Megen	25-12	21:00	08.30
Lith	25-12	22:00	06.32

The 1995 floods

Toward the end of January 1995, heavy rainfalls in France and the Ardennes raised anxiety in the province of Limburg. The memory of the 1993 floods was quite fresh and people were not willing to take any risks. On Monday 23 January, villages along the Meuse were taking the necessary measures to prepare themselves for a new flood.

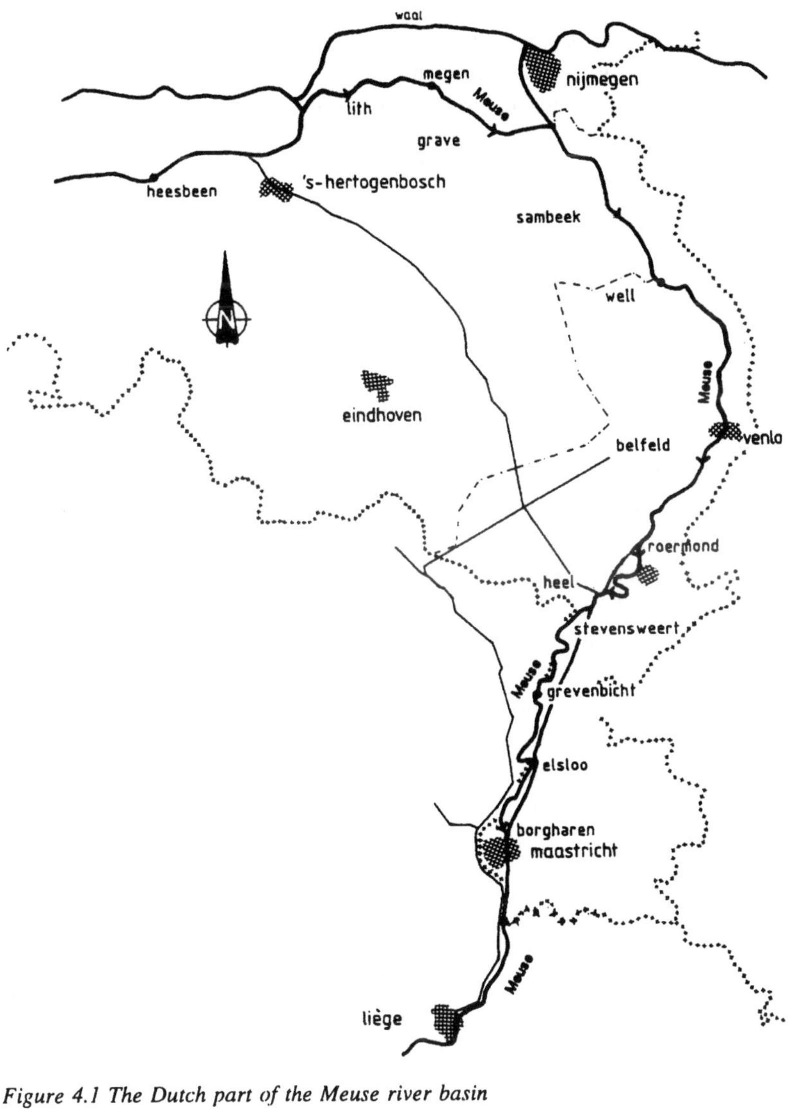

Figure 4.1 The Dutch part of the Meuse river basin

On Tuesday the water level was still under the height of the 1993 flood. Nevertheless, crisis centres were being set up in case a new flood might occur. On Wednesday 25 January RWS Limburg made an alarming prognosis for the coming days for the water level of the river Meuse. It seemed the water level was reaching a level that would exceed the one of 1993. This was the definitive sign for action. The mayor of Maastricht advised inhabitants of Itteren, Borgharen and parts of Maastricht to evacuate. Most inhabitants of these threatened areas choose to ignore these calls for evacuation. Ministers and members of parliament visited the province and showed their sympathy with the population.

In the course of the week, the situation deteriorated quickly. Inhabitants of Limburg were trying to save their belongings from the rising water. Every village along the Meuse had set up its own crisis centre. In several villages, the "state of emergency" was declared, enabling the mayor to force the inhabitants to leave their homes. On Friday 27 January, the provincial coordination centre reported that about 2,630 people had sofar been evacuated. More than 13,000 houses were flooded by the water.

Beyond the Meuse basin — Not only the river Meuse gave problems in 1995. The water in other major Dutch rivers such as the Rhine and the Waal, was also rising to alarming heights. This meant that provinces other than Limburg would also be at risk as well this time. Especially in Gelderland, the situation seemed alarming (see figure 4.2).

Part of Gelderland is situated between the rivers Rhine, Meuse, Waal, Lek and Lower-Rhine. The poor condition of some dikes in this region raises the risk of a breakthrough. In case the dikes burst, some of the low level polder land located between the dikes of the various rivers would be flooded with water within a few hours (the so-called 'bathtub effect'). This could quickly lead to a life-threatening situation for the inhabitants of this area.

Assessing the risk — On Saturday 28 January, the polder boards, responsible for the maintenance of the river dikes, declared that the situation in the "river area" was alarming.[4] They thought the prognosis of RWS Oost-Nederland for the Rhine was too optimistic and instead presumed that the situation would deteriorate during the weekend. They based their opinion on the situation in Germany where heavy rainfalls and melted snow caused the water level of the Rhine to rise further. In the evening, RWS updated its prognosis, confirming the opinion of the polder board. Meanwhile, preparations in the river area were being made in case an evacuation was necessary. The next day, Sunday 29 January, a crisis meeting was held in Arnhem between the provincial governor and the three coordinating mayors within the province of Gelderland, Nijmegen, Arnhem and Rivierenland. During this meeting, it was discussed whether an

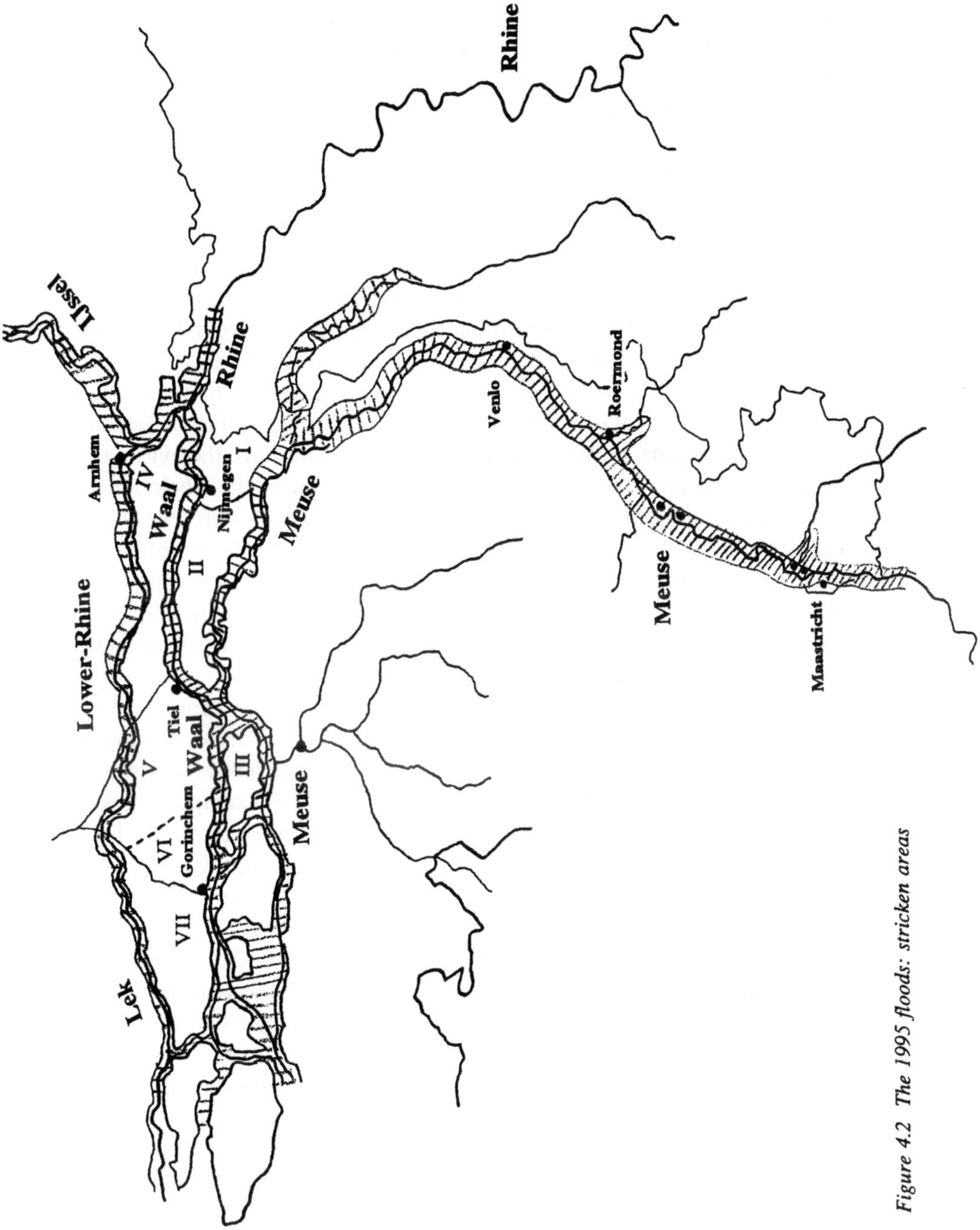

Figure 4.2 The 1995 floods: stricken areas

evacuation of the first part of the region (Ooijpolder and Land van Maas en Waal located in the area around Nijmegen) would be necessary. A representative of the polder board, in which this area was situated, stated that from Tuesday on the board would not take any responsibility for the safety of the dikes.[5] The participants all agreed that an evacuation was unavoidable. The final decision would be made in the regional coordination centre of Nijmegen in which all of the mayors primarily responsible for public safety and order in their municipality were represented.

Decision making about evacuation — On Monday 30 January, the regional management team in Nijmegen decided to evacuate part of the region. When the provincial governor advised the people of the region to leave their homes, most of them were already on their way. The alarming signals from mayors, representatives of polder boards and the provincial governor had caused a voluntary evasion by a great part of the population in this region. From Tuesday 31 January 9.00 onwards, nobody was allowed to enter or leave the evacuated area. Until that moment, everyone had the chance to move their belongings including cattle to safer places. Hospitals and nursing homes were evacuated on Monday. Assistance was provided from all over the country. In the following days it was decided at the provincial coordination centre to evacuate the other areas as well. In some cases an emergency evacuation had to take place because a sudden burst in the dike made a breach possible. Eventually, nearly 250,000 people had to leave their homes.

Problems occurred regarding farmers, the trade and industry. Special plans were not available and it seemed that the needed time for these groups to evacuate needed to be much longer. For example, for one corporation it would take about fifty days to move its stock to another place. Other examples showed that farmers and company directors were reluctant to leave their business. They declared that they would take the risk. The economic damage would be too high for them. Despite these problems, the evacuation went well.

The end of the evacuation — Even though the water level did not rise any further the situation remained critical. The period in which the water level was high had soaked the dikes. Therefore, the chance of a dikeburst was still substantial, but people were getting anxious to go back to their homes. The provincial governor and the minister of the Interior stated that this would not be allowed. On Thursday 2 February, rumours were heard that people might be allowed to go back the next weekend. The decision for this return would be made within a few days. At that moment the minister of the Interior decided to use his authority to intervene in the decision making process. Every decision concerning the return of the people should be discussed with him first.

The next day, the water level of the rivers declined rapidly. It was assumed that it would only take a few hours before the population could enter the evacuated areas again. Nevertheless, the decision was postponed until Saturday, since the area was not deemed safe enough by the authorities. A conflict arose concerning the timing of the message announcing the return. The provincial governor wanted to communicate the message as soon as the decision had been made, while the coordination centre in Nijmegen insisted to wait until the population could actually return. This would prevent chaos on the roads. In the end, it was decided that the people would be told the moment it was actually possible to return. In a press conference on Saturday morning the minister stated that the people from some evacuated areas could return to their homes. The return started on Saturday 4 and Sunday 5 February and went well.

Characteristics of the 1995 floods — Like the 1993 flood, the 1995 riverine flood had a gradual onset. However, due to the persisting period of rain, the 1995 differed from the 1993 flood. First, the 1995 riverine flood was characterized by an extremely long-drawn shape, whereas in 1993 extreme water levels held much shorter. Second, due to the extremely long duration of the flood, the high water levels starting at Roermond and further downstream were even higher than in 1993 (see tables 4.1 and 4.2). For days the constant transit of extreme quantities of water water-logged the soil in the winter-bed. Consequently, no extra water could be contained when peak-level was reached and the "high water-wave" hardly flattened as it usually does when a flood takes on a shorter-drawn shape (like the 1993 floods). Third, the long-drawn shape of the 1995 riverine flood caused a remarkable decrease in the period between the highest water level in Heel (South of Roermond) and Buggenum (North of Roermond) comparing 22 hours in 1993 versus only 3 hours in 1995. Table 4.2 (next page) represents the highest water level that was reached in sixteen municipalities (the same we presented in the tables 4.1 and 4.2), as well as the times and dates they were reached. Finally, the 1995 floods lasted longer: nearly two weeks as opposed to eight days in 1993.

The two floods compared

There were both similarities and differences between the two flooding episodes. The floods were similar in that they originated in the southeastern part of the country and gradually moved westward and both floods caused considerable damage but no deaths. Also, both floods had a slow onset typical of the Meuse and Rhine river basin. More important than these similarities were some key differences. First of all, the scope and impact of the 1995 floods were much bigger. They affected multiple rivers instead of only the Meuse and multiple provinces instead of only Limburg. A quarter of a million people were evacuated instead of 8,000. Also, the 1995 floods lasted longer: nearly two weeks as opposed to eight days in 1993. Moreover, the nature of the threat posed by the 1995 floods differed from their predecessor: in Gelderland, there was immediate threat to the lives and limbs of people and cattle, whereas in Limburg the type of flooding that occurs causes mostly discomfort. Lastly, whereas in 1993 most of the damage resulted from the impact of water on vehicles, houses, firms and public infrastructures, in 1995 a large part of the total damage was incurred by the economic costs of the precautionary evacuation.

Table 4.2 Highest water level of the river Meuse in 1995

Location	Passage highest water level Date	Time	Highest water level (+ NAP)
Borgharen-Dorp	31-01	5:00	45.17
Elsloo	31-01	6:00	40.23
Grevenbicht	31-01	16:00	32.73
Stevensweert	31-01	17:00	25.30
Heel-boven	31-01	21:00	22.69
Roermond-Buggenum	01-02	0:00	20.59
Belfeld	01-02	5:00	19.05
Venlo	01-02	7:00	18.46
Well	01-02	14:00	15.43
Sambeek	01-02	21:00	14.02
Grave	02-02	3:00	10.58
Megen	02-02	6:00	08.48
Lith	02-02	13:00	06.54

3 Risk communication

Context: disaster planning in the Netherlands

With major parts of its territory below sea level, flooding has always been a major threat to safety in the Netherlands. Dutch national history is indeed shaped in important ways by the country's battle against the sea. In modern times, the 1953 flood disaster wiping out 1/5 of the country's territory and causing the deaths of more than 1,800 people stands as a dramatic reminder of this fact. With that exception, Holland has long since been a safe country without many of the natural hazards (hurricanes, earthquakes, droughts) to be found elsewhere, and has experienced relatively few major disasters. However, industrialization, a fast-growing population and heavy urbanization have increased the salience of man-made risks and disasters, producing a series of major accidents (e.g. plane crashes, petrochemical explosions, train collisions). We begin with an explanation of the structure of Dutch disaster planning and response.

In 1985, the Dutch Parliament passed a new Disaster Act. This Act defines a disaster as an event that endangers the life and health of a large number of people, or causes severe harm to material interests, and requires coordinated efforts from various fields of expertise. This conceptualization of disaster reflects the prevailing principles of disaster management in the Netherlands. A very general notion of the social and economic disruption of the community is combined with an explicit demand for governmental activity and for coordination between governmental agencies. The Dutch emergency management structure is characterized by different levels of government being involved, i.e. central, provincial and local level.[6]

The Disaster Act has several provisions with regard to disaster planning procedures in the local municipalities. The Board of Mayor and Aldermen is charged with drawing up a mandatory disaster plan, which should describe the organizational structure and the emergency management activities as well as comprise a list of the agencies that may have to take part in emergency activities. The Municipal Council has to approve this plan. The mayor, who carries the primary responsibility for disaster response, is obliged to design disaster control plans. These plans should include a response scenario for possible calamities in the identified disaster-prone sites of the local community.

Local and regional fire brigades are the most important operational services involved in emergency planning. As the Disaster Act dictates, the local Fire Chief has the primary operational responsibility for the on-site coordination of local disaster responses. Generally, local emergency management coordination centres and operational centres are established. In addition to these centres, the so-called 'action centres' become active. Here teams composed of members from specific divisions perform tasks in fields including public relations, civil services, public works, environmental services (see figure 4.3).

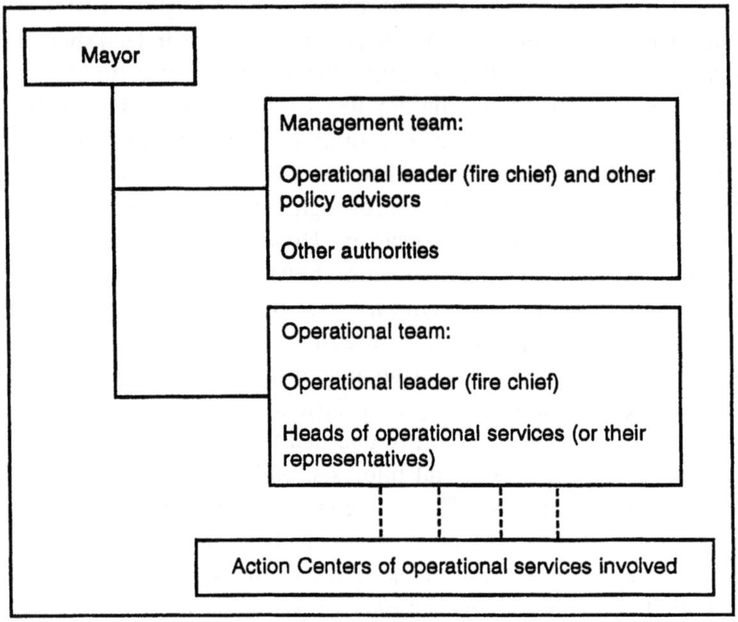

Figure 4.3 The organization of emergency management at the local level

In recent years, some executive agencies crucial to disaster management, such as the fire brigade and the police, have replaced a local mode of organization with a regional one. Unfortunately, the demarcation of these regions differs across agencies: the 25 police regions do not map neatly onto the 43 fire brigade regions. Nevertheless, this organizational transformation of operational agencies into regional bodies has required municipal authorities (in particular the mayors) to "regionalize" their approach as well. Moreover, when a disaster crosses the border of a municipality, regional coordination becomes necessary. At the regional level it may be decided to establish a Regional Coordination Centre, which supports the administrative and on-site coordination of emergency.[7] The Regional Coordination Centre is in charge of the information dissemination and functions as the communication channel between the local Crisis Centres and the provincial authorities. A mayor may be appointed as the administrative coordinator at the regional level without additional formal powers being delegated.

Provinces draw up their own provincial coordination plan. They have a formal coordinating role in the tuning of assistance demand and supply. Under the Disaster Act, the provincial governor has specific competencies which may be involved when problems cannot be solved at the regional or local level. The role of the provincial governor is embodied in various official documents such as the Official instruction, the Province Act, the Fire Act of 1985, the Police Act and the Act on Medical Aid in Disasters. Furthermore, this role is discussed in international treaties concerning international assistance.

In the preparedness phase of emergency management local authorities are often the key actors, following the essentially decentralized approach envisaged in the Act. When the situation becomes more serious and/or transcends the boundaries of one municipality, provincial or even national authorities (especially the Ministry of the Interior) may decide to coordinate or otherwise intervene. More specifically, when mayors do not cooperate regionally or when their approach runs counter supramunicipal interests, the provincial governor may give indications for administrative action. A similar rule applies between provincial governors and the minister of the Interior when a disaster transcends the boundaries of one province (see figure 4.4).[8]

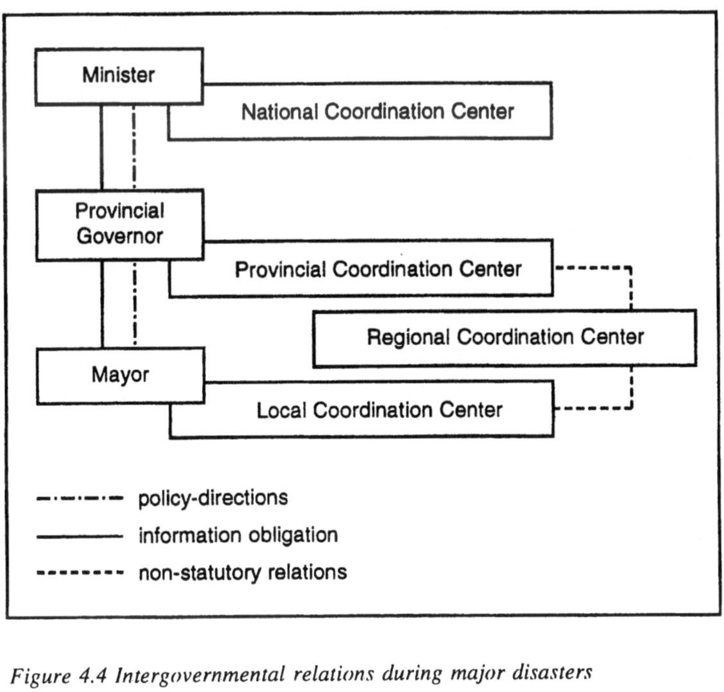

Figure 4.4 Intergovernmental relations during major disasters

The organization of intergovernmental warning

Several organizations play a central role in the dissemination of information pertaining to flood hazards. First of all, the Institute for Inland, Water Management and Waste Water Treatment (RIZA) predicts water discharges and water levels in both the Belgian and the Dutch part of the Meuse, using sophisticated modelling techniques.[9] The estimation of water discharges in the Meuse is based on the precipitation and discharge figures from Belgium and France, which are produced by the Belgian Office de la Navigation in Luik and the French Centre d'Annonces des Crues in Nancy.[10] In addition, the Ministère Wallon de l'Equipement et des Transport (Wallonian Ministry for Equipment and Transports) is contacted. Hence, RIZA's effectiveness is premised on optimal liaison arrangements with those agencies for an efficient and timely transmission of information. For the river Rhine, similar arrangements exist with the German Rhein Wasser- und Schiffartdirektion in Duisburg.

When the figures come in from the neighbouring countries, calculations and estimates are passed on by RIZA to the various regional directorates of the Directorate General of Public Works and Water Management (*Rijkswaterstaat*, RWS, part of the Ministry of Transport and Public Works). As far as the Meuse is concerned, RWS's involvement is structured by the 1985 'Plan for High Water in the Meuse'. Detailed estimates of the water discharge are made for the border village of Borgharen near Maastricht. Furthermore, estimates are produced for each municipality along the Meuse indicating the projected rise, the maximum height and peak-level times. In case of high water, these estimates are transmitted to the regional fire brigades in Limburg and Gelderland, as well as to the city halls of Maastricht and some other municipalities that have requested early warning information in the past.[11] RWS (Directorate General of Public Works and Water Management) predictions for the river Meuse are valid for a maximum period of 12 hours. Similar arrangements are in place for the river Rhine, albeit that the prediction period for the Rhine is much longer than for the Meuse (48 hours).

Other agencies involved in less prominent roles include the Royal Dutch Meteorological Institute (KNMI) and the Institute for the Coast and the Sea, also a part of Rijkswaterstaat. The entire warning and dissemination process is coordinated at RWS headquarters in the Hague.

Warning processes in 1993

As early as 14 December 1993, RWS issued the first early warning signal. This warning was withdrawn on 16 December 1993. At the same time, however, an excessive discharge was predicted for 20 December. On Sunday night, 19

December 1993, the Belgian TV station BRT first indicated the threat of a possible flood. That same afternoon the Belgian government warned that the water levels of the Meuse had risen to a critical height. At that moment the Limburg section of RWS was still unaware of the threat. Headquarters were closed and the municipality of Maastricht could not be reached. There was no contact between Belgian and Dutch authorities. A Belgian spokesmen said: 'No contact was made from the Dutch side. And we did not call them. This is not one of our tasks.'[12]

RWS came into action twelve hours after the first Belgian warnings. On the night of Monday 20 December, a second early warning was submitted to local, regional and central authorities. Starting this day, as the water level rose very quickly, regular reports on water levels and forecastings were issued. Furthermore, a high-water news service Meuse was started that same day. Information about the latest developments were available on this news service. Predictions of water level were transmitted once a day. Preceding the high-water news service, RWS Limburg provided for additional regional information to be sent to regional Fire Services in Limburg; district offices of the polder boards and other public authorities specifically dealing with in-land water management; regional police; its Flemish liaison agencies; the municipalities of Maastricht, Roermond, Mook and Middelaar (via the regional Fire Service); and the senior RWS directors in the Hague. In addition, RWS Limburg established an information centre in which information could be acquired about water levels.

From the moment the second warning was submitted, central, regional and local authorities became involved. RWS's messages were passed on by the regional Fire Services to municipalities.[13] When possible, the news service was supplemented with advice about measures to be taken. The available information on water levels was translated to the emergency stages in the (regional) disaster plans. In the southern region of Limburg and the region of Nijmegen, this had already taken place Sunday night. In Nijmegen, the mayors were contacted according to plan even though town-halls were closed at that time. In the central region of Limburg, municipalities were informed on Monday 20 December by the regional fire service, in the northern region on Tuesday.[14]

Regional differences — The region of *Southern Limburg* is always in the most vulnerable position, since its geographical location implies that it cannot do without a timely dissemination of information and warnings from neighbouring countries. Since in the 1993 case, contact between the Dutch and Belgian authorities was slow to develop and non-existent on the crucial first day, Southern Limburg was warned at a relatively late stage. Furthermore, it has the disadvantage that the river basin around Maastricht is quickly saturated when more rain water than usual comes in from Belgium. Hence changes in the volume of incoming water tend to manifest themselves rather quickly in the weather level in the Maastricht area.

In contrast to this high vulnerability, the sensitivity to risk appeared to be relatively low. Messages about the threatening flood were initially received with disbelief. The scepticism was enhanced by the experience of the past decade, when Southern Limburg had experienced several floods which all stopped short of causing major damage. The prevailing heuristic seemed to be that 'it will all blow over.' After Sunday night's warning, RWS decided to issue the next warning earlier as foreseen in the plan, because the water level rose very quickly (the second warning was submitted at 12.00 on 20 December and a few hours later it was clear that parts of Maastricht were going to be flooded by the Meuse).[15] In sharp contrast, however, the emergency services held on to their disaster plans, which dictated that operations should not start until a water level of 44.70m had been reached. No provisions were made for taking into account the remarkable speed with which the water level may rise. Consequently, they decided to refrain from warning the public about the coming flood. At 19.00, new RWS forecasts showed that the level of 44.70 would be reached soon and would increase even further. In response to this information, a municipal Coordination Centre in Maastricht was finally set up and the inhabitants were given a flood warning. On Monday evening 20 December and Tuesday 21, most southern municipalities started on a large scale to warn the public.

A further complication in the south was that whereas the municipalities were dependent on the messages from RWS for the implementation of measures, RWS could only provide 12-hour forecasts. Longer-term forecasts would lead to unacceptable margins of error, according to RWS's specialists. The southern municipalities, however, had little use for such short-notice forecasts, since these not provide enough time for adequate preparation. Furthermore, RWS forecasts appeared to be inconsistent, with predicted heights and peak-level times for one place differing with every new forecast. This increased ambiguity for local authorities that were looking for clear information, and who were unaware of the technical difficulties of estimation that RWS had to cope with.

The *Central* and *Northern Limburg* regions are less dependent upon the speed and efficiency of cross-national warning arrangements. In 1993, too, they had more time to prepare for, and inform the public about the coming threat, since it took about three to four days for the water to rise in these parts of the province. Nevertheless many of the communication problems that surrounded the dissemination of warnings by RWS also occurred in these regions (e.g. inconsistency of warnings, short-term forecasts). On the evening of Monday 20 December, the first warning signal was received by Venlo. The Regional Coordination Centre was not assembled until 21 December. Several measures were taken.[16] The first public warnings were issued on Tuesday (Roermond) and Wednesday (Venlo). The warning process in Central and Northern Limburg was further facilitated, paradoxically perhaps, by the confusion in Southern Limburg and its dire consequences: seeing the pictures of flooded streets and cars and a lot of avoidable damage in and around Maastricht, there was simply no scope for any of the authorities further downstream to play down the seriousness of the situation. Table 4.3 shows the warning times for the different regions and the measures they took.

Table 4.3 Inter-regional time lags in warning and response start-up

Municipality/Region	Early Warning		Regional crisis centre established/or central gathering at regional fire brigade centre if no regional crisis centre in place		Start active public information	
	date	time	date	time	date	time
Maastricht/Southern Limburg	20/12	2.00	20/12	15.30	20/12	19.00
Roermond/Central Limburg	20/12	2.00	21/12	6.00	21/12	12.00
Venlo/Northern Limburg	20/12	19.00	21/12	10.00	21/12	20.00

Warning processes in 1995

Limburg Towards the end of the third week of January 1995, the weather forecasts issued by KNMI predicted intense rainfalls for the weekend of 21 and 22 January. In combination with the effects of snow melting in the Belgian Ardennes, this was likely to cause flooding problems in the river Meuse. In response to the rising water levels, RWS Limburg set up an information centre in Maastricht, and starting Monday 23 January 00.30, the centre started its activities. At this time, active liaison with Belgian counterparts was taking place. At that time, water levels were expected to rise above 45m and the community of Borgharen near Maastricht, heavily affected in 1993, was under immediate threat of a new wave of flooding.

Interestingly, whereas in 1993 both the forecasting and response authorities did not initiate action until water levels matched disaster plan provisions, in 1995 they did not wait for that to happen and went ahead well before. Also in contrast to 1993, RWS was prepared to issue 'unofficial predictions' under embargo to mayors that contained a much wider time horizon than the official maximum of 12 hours.[17] On the basis of one such prediction, the so-called '46m plus scenario', early preparations were made for an eviction of Borgharen and Itteren, two small villages which are part of the municipality of Maastricht. Other mayors further upstream followed this pattern: they requested and obtained longer-range forecasts, enabling them to provide the agencies involved as well as the population with a better forewarning.

The more proactive attitude of authorities in the face of rising water levels was widespread, particularly in the south, where, as the 1993 experience had taught, the stakes were highest. This time also, the provincial governor played an important role early on. On 25 January, he assembled the coordinating mayors and regional fire commanders to share information and arrive at a common appraisal of the situation.

In contrast to 1993, the most important difference in regional attitudes and activities relating to warning was that this time around, Southern and Central Limburg were united in actively seeking information from the forecasters and from foreign sources, whereas the Northern region took more of a wait-and-see attitude. Presumably, the certainty of a slow onset provided a smaller incentive to more proactive information-seeking than the South's 1993 experience of surprise, confusion and avoidable damage. However, the uniqueness of the 1995 riverine floods in that it showed a remarkable decrease in time between the highest water level in Borgharen and Venlo (27 hours in 1995 as opposed to 37 hours in 1993) may have surprised the authorities in the Northern region. Another important difference appears if one compares the evaluations of RWS's information to local and regional authorities. In 1993, there was general dissatisfaction in all three regions, whereas in 1995 only the North's Regional Coordination Centre complained about the timing and usefulness of the forecasts. Also, the North was the only region to be confronted with media possessing the most recent forecasts before the Coordination Centre itself did. Apparently, a proactive attitude by "customers" facilitates a better connection between demand and supply of flood forecasts.

Gelderland — In a way, the warning process in Gelderland is both more complicated and simpler than its Limburg counterpart. It is simpler for the river Meuse because there is all of Limburg between Gelderland and the Belgian border. This reduces the chances of a "surprise flood" in this area. At the same time, the river Rhine tends to swell at a much slower rate than the Meuse, again reducing the risk of surprise through inattentiveness. On the other hand, the fact that Gelderland lies at the crossroads of various major rivers (e.g. the Rhine, Meuse and Waal) complicates the risk communication process, since there might be amplification effects. Also, Gelderland authorities need the cooperation of both RWS Limburg and RWS Oost-Nederland, increasing the burdens of coordination and information sharing.

The first warnings reached the regional fire service at Nijmegen by Monday 23 January (for the Meuse), and by 27 January for the river Waal, the southern fork of the Rhine, which divides just east of Nijmegen. This left the authorities with sufficient time to set up their response organizations and inform the public. One of the means of information set up was a toll-free public phone line providing local citizens with the latest flood information in their area. The downstream regions (Rivierenland and Arnhem) went through the same process with roughly 24 hours time lag.

Interestingly, problems occurred in the communication between RWS Oost-Nederland and other water authorities on the one hand (the providers of warnings and risk information), and regional response authorities (the consumers of risk information) on the other that were highly similar to those experienced in Limburg in 1993: response authorities wanted more long-range forecasts and specific, policy-oriented advice that water authorities were reluctant to give. For example, when the Rhine reached extreme levels in parts of Germany, civil authorities in Gelderland wanted RWS to revise its original forecasts, which did not take into account these developments. RWS however refused to issue new forecasts until it had all the information it needed and all the calculations had been conducted in the proper fashion. The clash between crisis management urgencies (in Gelderland, early information about water levels was essential to prepare for mass evacuations that might quickly become necessary if certain critical thresholds were passed – the 'bathtub effect') and professional norms and standards was evident here. Apparently, the experience gained by RWS Limburg with this dilemma – which had driven it towards a more pragmatic approach in 1995 and better cooperation with civil authorities – did not resonate in its neighbouring Oost-Nederland division.

Informing the public

In times of disaster, people generally expect prompt and accurate information. Warning at an early stage enables people to take necessary precautions and preparations to limit the effects of disaster as far as possible. The warning stage appears to be one of the crucial conditions determining the nature and outcomes of disasters. The timing, medium, and quality of warning information are essential.

The role of *timing* is straightforward. When disasters hit quickly and unexpectedly, governments will have little time to warn people. Consequently, the evacuation process tends to be more improvised, chaotic and less effective. The reverse holds true for disaster with a slower onset (and/or a more proactive warning regime). Given these general patterns, a key hypothesis of this study has been that the further downstream a community, the longer the warning time, and the more orderly the response process. As far as the *medium* is concerned, research has shown the main importance of the fact that messages originate from the authorities.[18] In natural disasters local authorities appear in a majority of the cases the first source of information, followed by family/neighbours/ friends (so-called 'informal networks'). The most important source of information in natural disasters are the mass media. Finally, the *quality* of the content of the warning is important. Citizens not only expect warnings to be timely, they also want warning messages to be as specific as possible about the nature of the threat and the kinds of countermeasures that are feasible.

To investigate the warning process, the Crisis Research Center conducted a survey among the inhabitants of some municipalities in Limburg affected by the floods of 1993.[19] Following the 1995 floods, a similar survey was conducted, covering both Limburg and Gelderland.[20] The surveys focused on people's experiences and assessments with regard to the provision of warnings and risk information; their responses to this information; their satisfaction with the information provided; as well as a number of other themes not related to the information theme. Below we describe the survey results focusing on the three core variables mentioned above. First we briefly introduce the general organization of the warning process.

The organization of the warning process — During both floods, information was communicated to residents through government sources and the media. The communication through government channels can be divided in "active" and "passive" information services. Warning is part of the active provision of information to the public. In 1993, for example, this active mode of government information provision was most essential in the initial stage, as the flood approached. Residents were encouraged to make preparations (turn of the gas and electricity and move valuable belongings to higher and safer places). Furthermore, this form of information was used to inform people how they could get subsequent information. At the same time, in both floods municipalities also provided a passive information service, reacting upon questions from the public, press or others. This passive form of public information became more important after the immediate threat had passed. Contrary to the coercive style of the active information, aimed to influence behaviour, passive information was advisory and left more room for citizen's discretion.

In sum, a variety of mediums was used to inform the public during both floods. Information about the approaching high-water was mainly received by the public via local authorities (flyers distributed among the public); regional/ local radio; regional television; police/fire service; information desks/toll-free numbers; family/neighbours/acquaintances. Below we shall present citizens' assessments of these various warning modes and information channels during both floods.

Onset and warning: regional differences — Compared to the other two parts of the province, warning time in Southern Limburg was shortest. We asked respondents when they were informed about the damage and risk that might result from the impact of the water in their residential area. The results showed that half of them were either not informed about the possible consequences of the flood until the flooding had started or not informed at all. Yet, a quarter of the respondents was aware of the coming water a day before the flooding. The hypothesis that the further downstream a community, the longer the warning time was partially confirmed: inhabitants of threatened areas in Central Limburg were warned at an earlier stage than South Limburg residents; however, in Northern Limburg inhabitants were generally informed at an even later stage than residents in Central Limburg.

Overall, the earlier respondents had been informed, the more positive their assessment of the timeliness of warning. In Central Limburg, residents were most positive. Whereas, in Southern Limburg, they were most negative (a majority said they were informed too late). Likewise in Northern Limburg, where residents were informed at a rather late stage, a majority of the respondents considered the moment of warning too late. Table 4.4 shows the percentages of respondents who were (very) content about the moment of warning.

Onset and warning differences: from 1993 to 1995 — In 1995, communities in Limburg were considerably less surprised by the rising water than in 1993. In every region the percentage of respondents content about the warning time increased, notably in Northern and Southern Limburg. In these regions nearly all respondents considered the official warnings to have been issued on time. Although, in Central Limburg in 1993, most respondents were already satisfied about the stage of warning, they were even more content with the timeliness of warning in 1995 (see table 4.4).

In 1995 signals that a life-threatening situation could occur, penetrated the Gelderland regions already at an early stage. Consequently, the authorities had enough time to warn the residents about the coming threat and a (possible) evacuation. Indeed, most interviewees acknowledged they warned (long) before the day the evacuation decision was made. Still, it is worth mentioning that still one out of every five respondents was not informed until the day of evacuation.

We also asked the respondents to indicate if they found the warning about the imminent evacuation well-timed or not. More than seventy percent of the respondent indicated that they were content, while thirty percent said the information came too late (see table 4.4). When the three regions respondents in Gelderland were more satisfied about the warning time than the respondents in Limburg in 1993 (except for the respondents in Central Limburg), however, they were less content than the respondents in Limburg in 1995.

Table 4.4 Timing of warnings: citizens' evaluations

Region	Respondents that found the evacuation well-timed (%)	Region	Respondents that found the evacation well-timed (%)
Southern Limburg 1993	25	Northern Limburg 1993	37
Southern Limburg 1995	98	Northern Limburg 1995	92
Central Limburg 1993	80	Rivierenland	71
Central Limburg 1995	92	Nijmegen	73

Use and evaluation of information sources — For the Limburg 1993 case, respondents were asked to indicate from whom or what they received the first information about the rising water. In Southern Limburg, no specific sender or means of transmission appeared to be dominant (fire services, loudspeaker vans, RWS and flyers to the public were mentioned). A notable amount of respondents (44 percent) received their first information from personal observation, or from neighbours and/or relatives. Part of the explanation for the low incidence of the municipality as the first source of information may be found in the fact that authorities were taken by surprise by the rapid rise of the water and consequently had not enough time to warn the public. In Central-Limburg, however, municipality was the most important supplier of information, as was the case in the North, albeit to a smaller extent (see table 4.5).

Throughout Limburg, municipalities used flyers to the public as a means of communication. In general, this dissemination mode worked well, particularly in Roermond where a large majority of the respondents was satisfied about the flyers (88 percent). In the southern region, where three quarters of the respondents read the flyers, a large proportion of the respondents was content as well. In Northern Limburg this means of (direct) municipal information was least effective. Although again, three quarters of the residents read the flyers, they were less positive than people in the two other regions. In some cases, their criticism was directed to the flyers' content: the information was said to be unclear, insufficient, or out of date.

During the 1995 floods in Limburg, the importance of mass media as a source of information increased strongly. Overall, approximately half of the population received their first information via the media. The largest percentage of respondents named television, followed by Radio Limburg. The percentage of respondents that mentioned the municipal information service as the first source equalled that of 1993. A lower, yet still significant, percentage of respondents, called acquaintances, family and/or neighbours, or personal observation as a first source of information (see table 4.5). The amount of people who read the official flyers decreased somewhat in comparison to 1993, yet their satisfaction with these flyers clearly increased, particularly in the Southern and Northern Limburg.

Table 4.5 Primary source of information in Limburg 1993 and 1995

Region	Source of Information			
	Mass Media	Municipality	Informal Network	Personal Observation
Southern Limburg 1993	14%	42%	19%	25%
Central Limburg 1993	20%	72%	2%	6%
Northern Limburg 1993	24%	47%	20%	9%
Limburg 1995	45%	35%	11%	9%

Also with regard to the 1995 floods, in the Nijmegen region most respondents called upon neighbours, relatives, friends or colleagues as their first source of information. This was less the case in the Rivierenland region. As in Limburg, mass media played a prominent role in the 1995 Gelderland floods. During the interviews we conducted, key decision makers repeatedly mentioned the importance of 'Omroep Gelderland'. This was confirmed in the population survey. In the Nijmegen region, 14 percent mentioned 'Omroep Gelderland' as first source of information and in Rivierenland region this rose to 23 percent. The authorities were aware of the effectiveness of local media, and used the radio station to inform the population. This partly explains why in the Gelderland regions, government agencies were called to a lesser extent as first source of information (see table 4.6).

Table 4.6 Primary source of information in Gelderland 1995

Region	Source of Information			
	Mass Media	Municipality	Informal Network	None
Rivierenland 1995	46%	22%	29%	3%
Nijmegen 1995	37%	22%	40%	1%

Evaluation public warning and information provision — To assess public opinion about the quality of warning and information provision, we will focus on two questions in the surveys related to these themes. First, we asked respondents which of the various sources of warning and disaster information they found most reliable. Second, we asked more specifically how citizens evaluated the accuracy of the information provided by municipal authorities in the various regions.

In the 1995 Southern Limburg case, most reliable information came from divergent services (e.g. municipality, crisis centres, media). In this region none of the organizations played a dominant role, whereas in Central Limburg the municipality of Roermond performed such a central function. Inhabitants received the most reliable information from this municipality. In Northern Limburg, local radio and regional television *(Kabelkrant)* were most valued (20 percent and 33 percent respectively), followed by the municipality (12 percent). In 1993, large differences existed in the satisfaction rates about the municipal information service. Central Limburg respondents were most satisfied. In Southern Limburg to the contrary, large dissatisfaction prevailed. The main reasons included a plain lack of information, incompleteness of information in the flyers to the public, and outdated information. In the northern region, almost half the respondents were content or extremely content, whereas approximately one third appeared to be dissatisfied or very dissatisfied. Overall, North Limburg residents underestimated the seriousness of the flood, due to insufficient information, inaccurate information or not having received any information at all. Other research supports the observation that the quality of public information provision was mixed, and left room for important improvements (table 4.7).[21]

During the 1995 floods, a majority of the Limburg respondents received most reliable information via the media (57 percent). One fourth mentioned the municipal information service as the most reliable source of information. In Southern Limburg inhabitants were most critical: only a relatively small portion said most reliable information came from the municipality. The bad experiences of 1993 may have lingered. In the other two regions trust in the municipal information service was larger (between 27 and 40 percent). In 1995 almost all Limburg respondents said they were warned on time. This inter-regional levelling is also evident in other parts of the survey. There was a slight decrease in satisfaction in Central-Limburg, while the percentage of (extremely) satisfied respondents in the remaining regions increased (table 4.7).

Mass media and local (emergency management) authorities were also important during the 1995 floods in Gelderland. Overall, half the residents named media sources as their reliable information. Once more the significance of 'Omroep Gelderland' for the process of information transference was emphasized: 30 percent singled it out. Over one third of the respondents (37 percent) put their confidence in the municipal information service. When compared to the score in the Nijmegen region, the score in Rivierenland region was higher (25 percent respectively 40 percent), which may be explained by the great success of the flyers to the public used in the Rivierenland region. Judgements about the quality of municipal information service corresponded with those about the moment of warning: respondents in Gelderland were more satisfied about the information service than the Limburg' respondents in 1993 (with the exception of respondents in Central Limburg in 1993), however they were less content than those in Limburg in 1995.

Table 4.7 Quality of municipal information provision

Region	Respondents that found the municipal information reliable (%)	Region	Respondents that found the municipal information reliable (%)
Southern Limburg 1993	32	Northern Limburg 1993	47
Southern Limburg 1995	76	Northern Limburg 1995	80
Central Limburg 1993	88	Rivierenland	66
Central Limburg 1995	75	Nijmegen	64

Media management

Media management in Limburg 1993 — Several agencies provided information to the press: province, RWS, Regional Coordination Centres/Regional Fire Services and municipalities. The province took both reactive and proactive actions to keep the media informed. They intended to take some work off the municipalities and regions. Moreover, the province coordinated the visits of the members of government and the media representatives.

An important part of news supply to the media was handled at the local and regional levels. In the northern region of Limburg, an important part of the information service to the media was managed at regional level, meaning that it was coordinated and directed from the centre municipality of this region, Venlo. In Central and Southern Limburg, municipalities handled the information service to the media, albeit often in a highly improvised and slightly disorganized fashion. Involvement from the Central Limburg coordination centre as well as the quasi-regional coordination centre at Maastricht was confined to advice to their local counterparts.

The media and the municipalities benefitted from each other's services. On the one hand, in discussing the information service to the public, we indicated that the local and regional media proved to be a particularly valuable medium for the municipalities. The media provided the mayors with an opportunity to showcast their involvement in the community's plight. This was especially so in those municipalities possessing a local radio or television station. In addition, local television and radio stations proved an invaluable means of communication with the public.

Despite the marked role of local media, a number of mayors indicated that in their view the most important image of the floods was created via national television, in particular through the news bulletins. On the national media scene, local authorities had a less conspicuous role, with the provincial governor acting as the chief spokesman for the inundated province. On the other hand, municipalities facilitated the media by the provision of information, technical facilities (such as telephone and fax) and catering services. Although media and municipalities worked closely together, the media performed its own role as an independent news service to the public. Thus, municipalities were not the only source of information for the media. Independent actions of the media could complicate the information service at local level. For instance, RWS provided information to the media. In some cases, the municipalities had not yet received this information. This could endanger the credibility of municipal information provision, since citizens and media alike were quick to pick up on these discrepancies.

Media management in Limburg 1995 — In 1995 a similar pattern of media management evolved. Again, the symbiosis between media and authorities was quite marked, with hardly any signs of conflict or scandal. Media reports proved a useful if sometimes confusing alternative to sluggish or incomplete bureaucratic information provision, such as the so-called Situation Reports (SITRAPS) and feedback from higher to lower levels in the response organization. Also, local authorities again used national media to draw attention to the plight of the stricken regions, and, occasionally, to put some pressure on national authorities to be forthcoming with financial support. This time however, authorities in Limburg could not monopolize the national media spotlight, since the floods moved on to Gelderland, and created much more drama there, with mass evacuations of people and cattle and the spectre of the 'bathtub effect' drawing major attention from national and international reporters.

A marked difference between 1993 and 1995 was noticeable in the professionalization of local-level media management. On the whole, better facilities were made available, and local authorities responded to media attention in a more coordinated way.

Media management in Gelderland 1995 — The beleaguered river dikes of Gelderland and the spectacle of townships threatened with instant inundation were impossible to resist for newscasters from all over the world. Local, regional and provincial authorities were besieged with media representatives. The different regions coped with this deluge in slightly different ways. In the Nijmegen region, for example, the press centre was kept strictly separate from the operational response centre and press contacts were centralized as much as possible in the hands of the municipal public relations officer of Nijmegen. Despite this arrangement, the overwhelming media interest was difficult to contain, particularly since at Nijmegen the first crucial evacuation decisions had to be made. Media management was much more problematic in the Rivierenland region, which did not have the means nor the facilities at its disposal to adequately meet the overwhelming media interest. Rivierenland's press centre was adjacent to the response centre, increasing the hectic atmosphere. Also, the sheer number of reporters caught the authorities unprepared. Media felt this too, and criticisms of Rivierenland's information policies were voiced. Finally at Arnhem, the media pressure on the regional coordination centre was perhaps slightly less strong, probably because there was never any evacuation. Media attention at Arnhem was directed towards the provincial coordination centre, where Governor Terlouw – once a high-profile party leader at the national level – proved an attractive and assertive spokesperson.

4 Disaster response

Key dimensions of the response that have been studied are the organization of the emergency decision making process and intergovernmental coordination.

The former focuses on how the formal disaster management structure operates in practice during the making of key response decisions. Coordination focuses on the ways in which the different administrative entities (local, regional, provincial, national) communicate and mutually adjust their responsibilities and activities in the response process. The case evidence is analyzed with respect to two themes: a. to identify the impact and limits of different formal coordination arrangements developed at the regional level; b. to examine the role of provincial and national authorities in the coordination of response efforts.

Disaster planning and flood management

Disaster control plans — Above, we have described the legal framework governing the organization of disaster response in the Netherlands. It became evident there, that the primary responsibility rests at the local level with the mayor. The involvement of regional, provincial, and national authorities follows a logic of subsidiarity. We shall come back to this in section 3.

The most important formal provision structuring the decision making process are the so-called disaster control plans that every municipality is supposed to develop to cover specific contingencies that may arise. All of these disaster control plans should fit the framework of the mandatory local disaster plan prescribed by the national Disaster Act of 1985. In other words, municipal authorities are supposed to engage in a risk analysis of their community, and develop a number of disaster control plans for accidents and disasters they deem most likely. These may range from petrochemical explosions to forest fires, and from crowd disasters to floods. The disaster control plan describes in detail the responsibilities and resources of various municipal agencies, and identifies at the outset the key decisional problems that may arise (decision scenarios), and how these are to be dealt with. In our case, we are especially interested in those disaster control plans dealing with responses to floods.

Since many disasters – almost by definition – transcend the boundaries of a municipality, a similar process of developing regional disaster control plans has been set in motion. The idea is that these regional plans should be complementary to the local ones and facilitate a smooth "upscaling" of disaster response should this become necessary; these regional plans tend to be the product of inter-municipal agreements. They arise from a "bottom-up" process, so to speak, since there is no regional level of government in the Netherlands at the present time, nor are provincial authorities legally obliged to initiate such regional planning. It will become apparent that there are great differences in the nature and extent of such bottom-up regional planning, producing a variety of de-facto decision making structures and processes during the two flood episodes. Some are primarily local, others regional, and in yet other regions the provincial level played a more prominent role.

Emergency planning in Limburg: 1993 and 1995 — Only the Northern and Central regions of Limburg had disaster control plans for water heights prior to the floods. These plans comprised both an organizational and an operational component. The disaster control plan of the Central region of Limburg contained an extensive description of the consequences of different water heights for every municipality and the actions to be taken by every organization involved. The disaster control plan of the Northern region of Limburg, on the contrary, was very brief. It did not provide local flood risk assessments and action plans; it only describes these in general terms.[22] In the absence of any regional plan in Southern Limburg, the only document pertaining to supramunicipal response in the region was the provincial coordination plan of Limburg. This plan provided a detailed elaboration of a. the duties and powers of the provincial governor in the event of a disaster; b. the reporting and warning procedures; and c. the organization of control and coordination, both in respect to the internal provincial organization and the rendering of assistance to other relevant organizations.

Following the experiences of the 1993 floods, Northern Limburg made significant changes in its disaster control plans. Specifically, the completely revised regional disaster control plan provided for a more clearly elaborated operational decision making structure. The coordination of operational agencies and activities was now delegated to a coordination team led by the regional fire chief and containing the heads of other regional services such as the police and medical services. In Central Limburg the already extensively and well developed disaster control plans were updated.

Remarkably, the 1993 experience did not prompt the development of a regional disaster control plan in Southern Limburg. It continued to rely on improvised modes of upscaling. One of the reasons why initiatives to develop a regional disaster control plan have failed was that three large cities in Southern Limburg (Maastricht, Heerlen and Sittard), strongly defended their autonomy. The provincial plan did undergo significant alterations. Specially, it created a provincial response team composed of the Governor and liaison officers from the operational services including the military, and supported by a variety of specialist advisory units.

Emergency planning in Gelderland 1995 — In 1993 Gelderland had been spared major flooding, but the water had risen to uncomfortable levels, highlighting the vulnerability of the existing river dike system and giving new force to advocates of dike reinforcement in their ongoing dispute with environmentalist groups who opposed such construction work. In the wake of the 1993 events, the Nijmegen region pressed for a more proactive emergency flood planning regime. Provincial and regional authorities agreed to appoint Nijmegen a pilot region for such planning efforts. This resulted eventually in the drafting of a master plan for regional response to floods. This plan was to be used throughout the province, thus including the Arnhem and Rivierenland regions. The plan mentioned four phases of emergency planning, related to specific water levels. Phases 0 and 1 entail preparatory activities, designed to facilitate the efficiency of subsequent mobilization and response efforts. Phase 2 starts when serious problems arise in one or several *dijkringen* (e.g. water regulation areas governed by a particular system of dikes and canals). In this phase, the regional coordination centre (RCC), supervised by the mayor of the region's principal city, is called into action. Phase 3 starts when there is a major threat and an actual disaster as defined by the 1985 Disaster Act has taken place. This gives mayors the authority to issue special by-laws, and at the same time strengthens the hand of supramunicipal coordinating bodies such as the regional centre. The plan was flanked by more detailed disaster control plans containing scenarios showing the consequences of dike failures in different places, as well as the evacuation measures that would need to taken in each of these cases. At the time the 1995 flood began to emerge, the new plan had just been adopted by the mayors of the Nijmegen region. It had been sent to the other regions as a model for their own planning efforts.

Crisis decision making: case experiences

In analysing the decision making process as it emerged during both floods, we have chosen to concentrate on evacuation, since this turned out to be a crucial issue in all regions, but particularly in Gelderland, e.g. the question of evacuation. In 1993, the decision making process concerning evacuation produced mainly negative decisions, e.g., decisions not to evacuate. In 1995, several large-scale evacuations were ordered, but some negative decisions were made as well.

Deciding about evacuation: Limburg 1993 — The surprise element that was so characteristic for the 1993 floods, particularly in Southern Limburg, also affected the strategic decision making process during the floods. Overall, decision making was primarily reactive, and more a product of haphazard improvisation to unforeseen and quickly changing circumstances than the enactment of well-grounded emergency planning. In fact, the existing plans either did not cover such a major flood contingency or had never been not taken very seriously by authorities basking in the comfort of reassuring historical analogies from the 1980 and 1984 floods (both of which had stopped short of causing any major damage). Consequently, decisions to evacuate tended merely to confirm what had already been set in motion by the rising water: they came at a time when the areas involved were either already inundated or on the verge of being so, leaving little scope for preparation and orderly evacuation on the part of local residents.

Overall, the dominant pattern of evacuation decision making was one of "situational dominance": ad-hoc and decentralized responses to compelling event-related imperatives.[23] The lack of effective planning was only one cause for this, the other being the reluctance of citizens to take the danger of flooding serious. They too were misled by comforting historical precedents of "inconsequential flooding", and were understandably reluctant to engage in precautionary evacuations.

This was especially so in Southern Limburg. In the other two regions, regional coordination centres were activated, but most of the attention from the authorities in those centres was focused on the situation in the principal cities in which they were located (e.g. Roermond and Venlo), leaving mayors from other towns largely to their own devices when it came to assessing the need for evacuation. An additional problem of the non-organized, essentially voluntary mode of organization that developed was that authorities were unable to exert much control over the pace and timing of citizens' return to their homes. This led to occasional tensions between citizens and officials, when the former wanted to be boated back to their houses, both during and after the flooding. There were also problems with people choosing to stay behind in their homes, and subsequently changing their minds or otherwise placing demands on already overburdened disaster assistance personnel.

Deciding about evacuation: Limburg 1995 — In the evacuation decision making process 14 months later, some important signs of social and organizational learning were evident. The reassuring analogies that had guided citizens and authorities alike, now met with competition from an all but reassuring analogy: the 1993 experience of "flooding *with* consequences". This produced a greater willingness in all parties concerned to appreciate the salience of worst-case scenarios, and to engage in more proactive planning and decision making. This was manifest in several ways. First of all, decisions to evacuate were taken much more early on in the flooding process. For example, whereas in 1993 the townships of Itteren and Borgharen near Maastricht had to evacuate by boats since no one wanted to consider leaving when there was still time to do so via roads, in 1995 the evacuation was initiated and completed more well before the water invaded the community.

In addition, a more sophisticated set of evacuation modalities was developed. Authorities in 1995 could chose from a repertoire of: (a) no evacuation; (b) encouraging a voluntary evacuation; (c) disseminating an urgent notice to evacuate (with or without a definite deadline); (d) declaring a compulsory evacuation. This provided them with a greater leeway, helped to prevent unnecessary anxiety, and facilitated compliance with evacuation orders once these were issued. In many cases, a large part of the people left and for that matter decision makers saw no need to declare a compulsory evacuation.

Also, particularly in Central and Northern Limburg, where regional plans existed and had been modified following the 1993 floods, there was a greater willingness to develop a coherent regional evacuation policy. In both regions, in the face of a relatively slow and benign development of the water levels, there was a deliberate regional agreement to hold off compulsory evacuation, yet leaving the ultimate descision up to the local mayors.

Finally, there was greater firmness on the part of authorities to convince people to evacuate. They were keen to minimize the number of people choosing to stay behind in their homes, and the problems disaster assistance personnel encountered that had hampered operations in 1993.

Deciding about evacuation: Gelderland 1995 — The impact of the emergency management planning in the Nijmegen region on the flood response did not remain confined to the written plans. The joint activities in the preparation of the plans appeared to be even more valuable in creating a better awareness of the risks involved, and the measures to be taken. They also helped to build networks between agencies. In fact, taken literally, the plans left a lot to be desired. First of all, their focus was almost completely on the decision making structure for a potential evacuation, and the technical means needed for its execution: Hardly any notice was given to the tasks, roles and responsibilities of the different authorities during an evacuation (e.g. informing residents of evacuated area, decision making about non-evacuees and/or remits for the evacuated area). Furthermore, the plans did not contain procedures to enhance regional administrative coordination, such as procedures to solve disagreement between local authorities (mayors) about the response and evacuation.

Despite these shortcomings, the joint efforts that had been made in the course of the Nijmegen planning process proved to be very beneficial. Authorities, along with regional and local operational services (fire service, police and municipal health services) had become involved in the planning process. Consequently, expectations and responsibilities with regard to the diverse emergency management tasks could more easily be attuned during the disaster response, resulting in time gains and efficiency. By virtue of its planning process, Nijmegen region was one step ahead of the other Gelderland regions. Whereas in Nijmegen region the emergency management services and authorities benefitted from the planning process, the other Gelderland regions lacked such experiences. They were compelled by circumstances to adopt the Nijmegen plans. The favourable position of Nijmegen was further enhanced by the fact that specific inundation scenarios were prepared, indicating the consequences of a dike burst and related to different time schemes. Whereas in the remaining Gelderland regions, such inundation scenarios were not always present.

The first evacuation decision was taken in Nijmegen, triggering a sequence of evacuations throughout the province. Ultimately, this process would lead to the evacuation of almost 250,000 people and hundreds of thousands of cattle. All of the twelve mayors in Nijmegen region were convinced that the decision to evacuate was inevitable. Their opinion was mainly determined by the steadfast position that the district in charge of a polder board Groot Maas and Waal took. The district announced no longer to guarantee the safety of the dikes, transferring their responsibility from the Groot Maas and Waal district to the local authorities, in effect pushing them towards making preparations for an evacuation.

Nijmegen region's decision to evacuate clearly set a precedent for other regions and provinces. First, it proved clearly that an evacuation could indeed be necessary and that the dangers of a dam burst should be taken seriously. Second, citizens' attention was drawn to the fact that the river dikes were not invincible. Following the Nijmegen evacuation, inhabitants of other regions in Gelderland started to question the quality of the river dikes in their areas. In some cases residents started to evacuate spontaneously. Third, some decision makers in other regions found it hard to accept that an evacuation in their area would not become necessary. The word was out that it was best to play it safe. Fourth, the success of the evacuation operation in Nijmegen encouraged the other regions to overcome their reluctance. Finally, preparations for the (possible) evacuation of other regions had already started, and gained momentum at the operational level. Disaster assistance personnel had gone out of their way to make the necessary preparations for an evacuation. Decision makers realized that a negative evacuation decision could frustrate these grass roots workers. In combination these factors put pressure on the mayors in the downstream areas of Nijmegen region, creating a snowball effect. After the evacuation of the Ooijpolder and the Land of Maas and Waal, followed the Bommelerwaard, the Tieler- and Culemborgerwaard and parts of the Betuwe, including the emergency evacuation of Ochten. The evacuation decision of Tuindorp (Arnhem region) was eventually reversed.

The advice or conclusions of polder board authorities were of great importance for the evacuation decision making processes. In some cases additional experts were called in. Although decision makers employed also other criteria, like rainfall figures and weather forecasts, the water authorities' recommendation tended to be decisive. However, large differences existed in the nature and quality of these recommendations as well as their phrasing. Differences also existed in jargon between the decision making (regional) authorities on the one hand and the water-authorities on the other hand, in some cases causing confusion about the significance of the information. For example, the polder board of the Betuwe region send an identical fax to the Regional Coordination Centres of Rivierenland and Arnhem. On the basis of the same information, both centres came to a different risk assessment.

The sequence of decisions to evacuate (the so-called 'evacuation bandwagon') ended in the province of Zuid-Holland. In close consultation with the province, local authorities in the Alblasserwaard and Vijfheerenlanden area came to the important decision not to evacuate this part of Zuid-Holland. This negative evacuation decision was remarkable, because the risks of a dike burst in this area were as big as they were in the evacuated Gelderland regions. The decision not to evacuate was based upon the different premises Gelderland and Zuid-Holland authorities used to assess threat (e.g. differences in dike heights (falls) and quality, and the capacity of exit routes). As in the Gelderland regions, advise of the water-authorities was decisive. The dike reeves considered an evacuation not necessary, and convinced all of the mayors in the Alblasserwaard and Vijfheerenlanden area of the safety of the dikes. The mayors also based their decision on the fact that an evacuation of this area would require a difficult, if not impossible, logistical operation (a so-called 'mission impossible'). Whereas in the Gelderland regions, residents were given time to take the necessary preparations and leave the threatened area, the authorities employed another scenario in the province of Zuid-Holland. In case the dikes burst, the residents would have to leave the area immediately because a serious life-threatening situation could occur instantaneously (the so-called 'doom scenario'). The question remains if inhabitants, in case of a dike burst, would have been prepared to leave their homes at such short notice, leaving behind all their belongings.

Deciding to return — Soon after the last evacuation, rumours about an imminent return quickly spread. On the one hand, the water levels suddenly dropped and the dikes did not break, hence citizens presumed that they could quickly return to their homes. The authorities, on the other hand, considered the areas not safe enough yet, denying residents access to their homes. As time passed, traffic on the exit-roads expanded and citizens' pledges to return became more insistent.

The most important criteria for the return decision were: RWS forecasts, advice of the polder board and independent experts.[24] There was a remarkable difference between the decision to evacuate and the decision to return. In the latter several other criteria were applied. Not only the economic but also the societal costs of keeping the evacuated areas closed were considered. Industry and farming representatives increased the pressure to allow a quick return.

Intergovernmental coordination and flood response

When a disaster transcends the boundaries of a municipality, the activities and capacities of the municipalities involved need to be coordinated. Moreover, the mayors may join their efforts in response to the disaster and thus, regional coordination becomes optional. Inter-regional differences exist in the inter-municipal initiatives and agreements about regional coordination and decision making competencies at regional level. Therefore, intergovernmental coordination structures between local and regional differ. A mayor may be appointed as the administrative coordinator at the regional level. This 'coordinating mayor' may be in the chair of a day-to-day joint meeting of the mayors in a specific region, or act as mediator between mayors on an ad-hoc basis. Yet, in other regions there may very well be no appointed administrative coordinator at all. However, it is important to note, that whatever regional construction may exist, no additional formal powers are delegated to an administrative coordinator. Thus, the primary responsibility rests at the local level with the mayor under all circumstances.

Whenever a local disaster affects supra-municipal interests, the provincial governor has the authority to coordinate and control the emergency management operation. These competencies, however, have been restricted as far as possible: the provincial governor may give directions on the policy to be adopted and may perform a supervisory role at the provincial level. Nevertheless, it is left to the mayors to decide, albeit on the strength of the policy directions, what the actual course of action is going to be, and how they put their competencies to use. Consequently, the mayors remain politically responsible for their decisions, which shows again the essentially decentralized approach envisaged in the 1985 Disaster Act. When a situation becomes even more serious and national interests are at stake, the Minister of the Interior may decide to give directions to the provincial governor on the subject of the (operational) emergency management policy to be adopted.

Upscaling — Whenever during a disaster existing administrative and operational response capacities do not suffice, organizations will alter their internal organization to enlarge their capacities. Every organization involved, – operational services, governmental agencies at local, provincial and national level, and private organizations such as the Red Cross – will restructure their internal organization to be ready for a large-scale operation (the upscaling process). The fire brigade, for example, will replace its local mode of organization by a regional one of companies or battalions. As for the governmental agencies concerned, this upscaling process implies that, depending on the scope of a disaster, the intergovernmental coordination structure, explained in the previous section will apply.

During the process of upscaling it may be necessary to allocate extra means and personnel to the threatened area. Hence, procedures and agreements regulate the assistance provided for by the police, the fire brigade and the various other counter-disaster agencies, as well as assistance from private organizations and neighbouring municipalities. Whenever, additional assistance is needed, local agencies need to follow the following formal procedure. The local requests for assistance need to be directed either indirectly, via the regional coordination centre, or directly to the provincial coordination centre. At the provincial coordination centre, the necessity of the assistance requests will be determined and an inventory of the possibilities within the province will be made. The provincial coordination centre needs to keep an overview of the assistance rendered and, if necessary, may rearrange the assistance provided within the province. Whenever during a disaster existing administrative response capacities do not suffice, the assistance of the military may be requested. The Provincial Military Commander may second units to the regional fire chief or the mayor for counter-disaster assistance, he stays responsible for their conduct and operations.

Should this assistance still not suffice, through the intervention of the Minister of the Interior, assistance potential from other provinces may be called upon. Moreover, Belgium, Germany and the Netherlands have committed themselves to render each other assistance to the best of their abilities.

Intergovernmental coordination in flood response: case experiences

Coordination in Limburg 1993: coping with the unexpected — The atypical and overwhelming scope and impact that was so characteristic for the 1993 floods affected the coordination process during the crisis. In what would turn out to be the largest emergency operation since the 1953 floods, a complex variety of local, regional and provincial coordination centres was activated. Apart from the fact that the existing disaster control plans did not cover such a major flood contingency, the coordination structures covered in the plans were never tested nor activated on such a large scale in the province of Limburg. Consequently, unexpected imperfections of the intergovernmental structure impeded the coordination process. Firstly, the processes for regional support and coordination were insufficiently known at the local level (e.g. which local counter-disaster organizations need to be represented at the regional level).[25] Secondly, some local coordination centres were understaffed or even non-existent (e.g. Roermond and Venlo), because their key decision makers had to attend to regional emergency management tasks. In other cases officers had to perform both local and regional tasks simultaneously, causing confusion and excessive workloads for those involved. Responsibilities and competencies of local and regional coordination centres became entangled. As a result, the authorities in the regional coordination centres were repeatedly accused of a biased preoccupation with the interests of the principal cities in which they were located.

In Southern Limburg, where regional coordination was almost non-existent, the situation was somewhat different. The regional fire brigade is supposed offer first-line assistance to the municipality of Maastricht. Consequently, regional officers acted as commanders for the local coordination centre in Maastricht, directing their main attention to the needs of this municipality. The provincial coordination centre tried to compensate the lack of regional coordination to the best of its possibilities. However, this did not prevent the smaller communities from complaining about the fact that assistance provided regionally was mainly directed to Maastricht.

Overall, the overwhelming need for rapid assistance induced municipalities to deviate from formal procedures, e.g., contacting non-neighbouring municipalities in Limburg or other provinces, foreign municipalities and agencies, and the Provincial Military Coordination Centre. Consequently, coordinators at the provincial level could never obtain a clear overview of where all the assistance had gone. This obstructed their activities. Regional and local coordination centres were also reluctant to send situation reports on a regular basis.

Due to large demand for support, the assistance of the military was requested. The military support turned out to be an integral part of the overall relief effort during the flood in 1993. It had been a long time since military services were used for counter-disaster tasks, the last major operation being the response to the 1953 flood in the southwestern part of the country. The conspicuous involvement of military personnel combined with the virtual absence of pre-existing civil-military coordination structures were a cause of problems and misunderstandings. Technically, the mayors were in charge of the response to the disaster in their municipalities, but this did not cover the role of the military. The military stuck to its own command structure, and liaison arrangements were not always working properly. In some cases, military commanders stayed away from the inter-service coordination meetings held at municipal and regional crisis centres. This led to some surprises and friction surrounding the rotation and withdrawal of military personnel in the disaster area. Civilian authorities had hoped for a much more protracted deployment of soldiers to assist in the recovery operations, but found out to their dismay that the soldiers were being pulled back by their commanders.[26] In another case, shortly after being refused a request for military assistance, a mayor was surprised to see soldiers appearing in his town after all.

Coordination in Limburg 1995: learning from failure — The coordination process in 1995 showed important signs of improvement. Some of the 1993 coordination problems, however, proved persistent. Overall, the coordination process improved because most of the counter-disaster personnel shared the 1993 experience. In Northern and Central Limburg, local and regional responsibilities, competencies, tasks and interests were more clearly segregated, although there was still room for improvement. Moreover, authorities were better acquainted with the formal coordination structure and the imperfections of the regional mode of organization. Authorities in the regional coordination centres did learn to divide their attention among all of the stricken municipalities, resulting in a considerable decline in complaints from local authorities about a biased preoccupation with the needs of the principal cities in which these centres were located. Also, the problems posed by the local representation at the regional coordination centres diminished. The problems of understaffing, however, appeared to be rather persistent, especially for the smaller communities.

Overall, intergovernmental requests for assistance were more structured, resulting in a considerable decrease in informal by-passes. This was partly due to the agreements made between the provincial, regional and local coordination centres and partly as a result of the guidelines send by the provincial coordination centre at the start of the floods. One of the more intractable problems, however, was the delay in time caused by lengthy formal procedures for assistance.

In Southern Limburg, the non-organized, improvised regional mode of organization that came into existence posed an additional problem. Analogues to the 1993 procedures, the local authorities initially contacted the provincial coordination centre for assistance. Unexpectedly, the local authorities had to direct their request to the regional authorities, who were not familiar with the existing formal procedures for support. Moreover, the regional authorities were criticized because they decided on the priority of assistance requests, a task that is formally a provincial one.

Overall, the process of cross-agency communication improved due to better infrastructural means for communication and the use of standardized SITRAP formats. According to several counter-disaster organizations and coordination centres at all levels, the SITRAPS were significantly weakening the speed of the immediate relief operation. Some coordination centres were overtaxed by the volume of information received (the Central Limburg regional coordination centre received over 1,970 faxes), yet others lacked information. Especially, the regional and provincial coordination centres gave no sufficient feedback to the lower levels of administration. Others complained about that requestsfor situation reports further exacerbated, the already high, workloads. Moreover, coordination centres at all levels of administration (local, regional and provincial) lacked an overview. At the local level insufficient feedback from the regional and provincial coordination centres was reason to complain.

Finally, the military service was one of the organizations that most clearly learned from the 1993 experience. Overall, local, regional and provincial authorities praised the liaison-officers at the regional and provincial coordination centres.

Coordination in Gelderland 1995: harmony and friction — Since, in Gelderland, a large number of municipalities were confronted with a possible need to evacuate, all of the mayors involved felt the need for regional coordination. In both Nijmegen and Rivierenland region, mayors gathered to discuss the possibility of an evacuation. In retrospect, the joint policy assessment process was generally valued a successful one.[27] Yet the underlying intergovernmental coordination mode producing the various evacuation decisions differed. We can distinguish at least five different coordination forms. They are as follows:

1 *Regional decision.* The decision to evacuate the Ooijpolder and the Land of Maas and Waal was first made at a joint meeting of all of the mayors at the regional coordination centre in Nijmegen. Next, the coordinating mayor of the Nijmegen region notified the authorities at the provincial level. The governor, then, informed the public at large.
2 *Provincial decision.* The decision to evacuate the Bommelerwaard was taken during a meeting at the Gelderland provincial coordination centre and was made public by the governor. Subsequently, nothing else was left to do for the coordinating mayor but to notify all of the Bommelerwaard mayors.
3 *Local primacy* The decision making process about the evacuation of the village Tuindorp (part of the Rijnwaarden municipality in the Arnhem region), somewhat resembled the decision making of the Bommelerwaard. Although, the mayors involved previously discussed the situation and agreed upon the premises for a possible evacuation, the decision to evacuate the village of Tuindorp was taken at the provincial coordination centre, without consultation with the responsible Mayor of Rijnwaarden. Subsequently the media were informed, causing the responsible mayor to be unpleasantly surprised when he learned about the decision via the radio. In the end, however, he proved able to reverse the decision.
4 *Authorization.* In the Rivierenland region, during a meeting at the provincial coordination centre, the mayors authorized the coordinating mayor to decide

on their behalf about the evacuation of the Tieler and Culumborgerwaard during a meeting at the provincial coordination centre. They agreed to be notified afterwards.

5 *Local decision.* The emergent evacuation of Ochten, part of the municipality of Echteld, was taken by its mayor. It was officially confirmed by the provincial coordination centre afterwards.

Despite these coordination differences, the primary legal responsibility for the decision to evacuate rested with the mayors. In cases of conflict or disagreement the mayors held a strong position. As to events unfolded, however, there was a growing need for coordination between the different regions and provinces. This was met with the expanding role of the Gelderland provincial coordination centre. The coordinating mayors of the Nijmegen, Rivierenland and Arnhem region gathered daily under the supervision of the governor. These meetings were meant to attune the (evacuation) decisions taken at various places.

Overall, the participants in these joint meetings thought they were useful. Nevertheless, some of them questioned the governors' role in announcing the evacuation decisions to the public. The fact that they were proclaimed at press conferences directly held after a joint meeting, created the impression that the decision was made by the governor himself, and making it seem as if the governor had overtaken the mayors' responsibilities.[28]

Secondly, some of the participants indicated that the rather vague status of the joint meetings at the provincial level was confusing. Partly, this resulted from the unspecified role and position of the coordinating mayors in these assemblies. Moreover, their positions differed among each another. Some wanted a joint agreement before they operated on behalf of the mayors in their region (e.g. Nijmegen), while others were authorized (e.g. Rivierenland).

Thirdly, the positions of the coordinating mayors in respect of one another were indistinct. During the decision making process at the provincial level, no clear distinction was made between the weighing of regional and inter-regional interests.[29] Consequently, coordinating mayors could interfere with the decisions to evacuate other regions. As a result, the coordinating mayor of Nijmegen region played a rather dominant role in evacuation decision making concerning the Tieler and Culumborgerwaard in the Rivierenland region, an area over which he had no jurisdiction whatsoever.

Finally, the coordination process was complicated by the demarcation of regions differed across agencies: the boundaries of administrative entities (municipalities, regions, provinces and polder board districts) and several executive agencies (police and fire brigade) did not match well.[30]

There was a creeping process of administrative upscaling. Whereas the first decision in Nijmegen region was taken at the regional level, subsequent decisions were de-facto taken at the provincial level. Ultimately, even the Minister of the Interior involved himself in the decision making process about the return to normalcy. The intervention of higher levels of administration was not generally appreciated. In the aftermath of the 1995 floods, criticism was levelled against the interventionist style of the governor and the Minister of the Interior.[31] Local actors throughout Gelderland openly doubted the effectiveness of this centralization of decision making. The centralized mode of crisis response alledgedly decreased the flexibility and swiftness of the operations.

5 Damage compensation

Damage compensation in the Netherlands

Shortly after evacuees returned to their properties, public attention shifted towards the assessment of damage and the question of damage compensation. This issue attracted much public attention since compensation for flood damages is not covered by private insurance companies in the Netherlands. Victims of the floods turned to the government and its National Disaster Fund to claim reimbursement.

The financial compensation of losses as a result of natural disasters has not been subject to formal regulations in the Netherlands prior to the 1995 floods. The Disaster Act covers only costs of emergency aid by municipalities. The discussion about the insurability of natural risks started after the 1953 floods in the southwestern part of the Netherlands. During this flood, 1,835 people died and more than 200,000 pieces of cattle drowned. The flood damage rose to 152 million guilders to buildings, 280 million guilders to farmland and 300 million guilders for the repairment of dikes.[32] The considerable damage was only partially compensated for by insurance companies after the 1953 floods. However, this disaster raised awareness of the risk of flooding and potential financial consequences. The Association of fire insurance companies conducted a preliminary study of possibilities for covering risk of flooding in current policies. This study showed that a disaster of the same scale as the 1953 floods would lead to a financial catastrophe for the insurance companies involved. The Association therefore decided to forbid its members to insure the risk of flooding. There were other arguments against insuring private properties for natural risks. Natural disasters can lead to large damages in upredictable ways. This places a great burden on the financial capacity of the insurers. Another reason had to do with the geographical concentration of risks. Most likely, only people in high-risk areas will apply for private disaster insurance. Since the risks cannot be spread among a larger population, premiums on such policies are likely to be high. Low income households in high risk areas may then decide to forego insurance, and will be left uncovered in case of a disaster.

This situation continued until the late seventies. In 1980, a commission was installed that reconsidered the possibility of including the risk of flooding in insurance policies. The central conclusion was that although there was a theoretically minimal chance of flooding, the consequences of different flood scenarios were exceedingly difficult to estimate. The results of this study gave no reason to change the no-insurance policy of the branch organisation.

Meanwhile, government was a minor player in all of this. The absence of major natural disasters since 1953 reduced public and political attention for the issue of damage compensation. In april 1992, the Roermond region in Limburg was hit by an earthquake. Total damage amounted to about DFL 100 million. Local and provincial authorities turned to the cabinet for financial assistance, which was initially reluctant to be forthcoming. It decided to pay up only after sustained public pressure and media inquiries. In the wake of that controversy, the discussion about the responsibilities of private and public actors in insuring citizens against natural risks was reopened. Regular meetings were now held between delegates from the ministry of Finance and the Interior and the Association of insurance companies. By the time the 1993 floods occurred, the parties had not reached any agreement, however. The issue of burden sharing in disaster risk insurance remianed deadlocked well after 1993, and by 1995 no new system was in place yet.

Damage compensation following the 1993 and 1995 floods

"Wet" versus "dry" damage — The 1993 and 1995 floods produced considerable damage. The situation in 1993 was relatively straightforward. Nearly all the victims in the Limburg region suffered damage due to the flooding of their properties ("wet damage"). Moreover, considerable costs were made to remove the debris from the houses and buildings. Nearly 6,000 people in this province and a smaller part (350) from other provinces experienced some degree of damage to their houses and furniture.[33] This added up to a total cost of DFL 254 million. Part of this damage was suffered by businesses (DFL 74 million) and farmers (DFL 19 million).[34]

The damage in 1995 was different from 1993 in several ways. The larger scale of the 1995 flood was a major factor as well as its longer duration. Nevertheless, the total damage in the province of Limburg was relatively smaller than in 1993. The experience of 1993 and the timely warning in 1995 enabled people to undertake the necessary precautionary measures. In the end, approximately DFL 350 million was paid by the government.[35] The main difference between the 1993 and 1995 floods concerned the type of damage. In 1993, most of the damage resulted from property being flooded. In 1995 Gelderland province eventually turned out not to be flooded, but it did see a large-scale precautionary evacuation operation. Considerable expenses were made to move belongings or assets from houses or businesses. Furthermore, businesses and farmers faced huge losses because they could not produce goods or deliver services. The main problem was that these costs, referred to as dry damage, were difficult to assess, posing a major problem for the organizations dealing with damage settlement.

Indemnities: negotiations and settlements — After the 1993 and 1995 floods, a public debate was held about liability and damage compensation. With private insurance options lacking altogether, victims were anxious to know who would compensate for the damage. Most claims were addressed to the national government and were supported by several members of parliament. In the absence of private insurance, the government was held responsible for damage compensation. With the water still running high, the cabinet announced that it would compensate the damage caused by the flood: 'we will not be mopes. If we have to pay, we will not be childish.'[36] Eventually, this resulted in negotiations between the government and representatives of different parties. Several victims applied for compensation: citizens, business, farmers, the public sector and some smaller parties. In this context, the public sector refers to those sections of government which faced damage to public buildings and roads, and the other parties to organizations and institutions such as hospitals, churches and associations and foundations. These victims all faced damage and costs because they were located in the flooded or threatened area.

Compensation in 1993 — In consultation between the province of Limburg, the National Disaster Fund, the Dutch Red Cross and the departments of the Interior and the Finance Department it was decided that only non-insurable damage of private *citizens* would be taken in consideration.[37] The settlement was set up according to earlier compensation measures taken with regard to the 1992 earthquake.[38]

Provincial authorities in Limburg initiated the so-called Floods Foundation December 1993, tasked with the inventory of damage to premises and furniture and the implementation of the settlements. The National Disaster Fund controlled the available funds. This led to the following settlement: victims whose damage was less than DFL 20.000 received 90% compensation; for the next DFL 10.000, victims received 75% compensation; and for any additional DFL 10.000 of damage, 50% compensation was awarded.[39]

Shortly after the floods, it was rumoured that the flood would bring many *businesses* near bankruptcy. The minister for Economic Affairs, responding to wide ranging public feelings, declared that 'no business will be allowed to go bankrupt due to the flood.' This meant that businesses would be eligible for considerable financial assistance. During implementation, however, there was a lack of organization. Different organizations representing the local economic interests were eager to act as spokesmen for victimized businesses in the region, creating considerable confusion. In the end, a foundation similar to the one set up for paying compensation to individual households was started. The participants in this Floods Foundation Corporations Limburg were the Chamber of Commerce, the Institute for small and medium-sized businesses and the Organization for Independent businesses in Limburg. This foundation negotiated with the department of Economic Affairs about a possible settlement. The chairman of the foundation was a former Economic Affairs minister, which was obviously an advantage in the negotiations with the department. In the absence of any pre-existing departmental policy, the foundation obtained considerable leverage.

Two main problems arose. The first area of disagreement concerned the size of the responsibility that businesses themselves would take on. The minister felt that a 50/50 division between the government and the businesses was appropriate. The members of the foundation disagreed and turned down his proposal. Only after strong pressure from several members from parliament was the minister prepared to offer a more generous settlement.[40] A second problem dealt with indirect damage. Many businesses experienced revenue losses during the disaster period. A contributing factor was the fact that the flood occurred between Christmas and New Year, the busiest period in the year for businesses such as shops, hotels, and restaurants. These indirect costs were difficult to asses, however. Revenue losses were eventually excluded from the settlement. However, businesses could apply for a gift from a special fund when their solvability was threatened as a result of indirect flood damage. In the end, the following settlement was agreed to: (a) 65% compensation of the damage to assets; (b) a special credit settlement (favourable conditions for loans); (c) a special fund for businesses with major problems.[41]

The procedure for the *farmers* was nearly the same as for businesses. The only difference was that in the past year farmers in other parts of the Netherlands had been confronted with damage to their property due to bad weather conditions. In the case of the 1993 flood the minister of Agriculture declared that financial assistance would be provided to farmers who experienced financial problems. The indirect effect was that farmers from other regions in the Netherlands asked for financial assistance despite not having been flood victims. A foundation was established to implement the settlement. Negotiations between the foundation and the department of Agriculture led to a settlement where 65%

of the damage suffered was compensated and special credit settlements were offered.[42]

Compensation in 1995 — The compensation process for 1993 was just finished when the 1995 flood hit the Netherlands. Again victims turned to the central government for damage compensation. The ongoing negotiations between the central government and the insurance companies had not yet resulted in an agreement. Prime minister Kok declared that the government would compensate the damage caused by the flood of 1995.[43] Most of the settlements for 1995 were based upon the experience of 1993.

Several settlements applied to the compensation of the damage of *citizens*. The National Disaster Fund received an amount of DFL 82 million which was subsequently doubled by the national government. This fund reimbursed 90% of the damage to buildings and between 50% and 100% of the damage to the furniture. The ministry of the Interior issued several arrangements for the compensation of the damage, the most important being a DFL 500 settlement for evacuated households. In the autumn of 1995, a complementary fund was set up. Furthermore, contributions were made to *municipalities* according to the Disaster Act. Their costs of emergency aid were reimbursed by the ministry on the base of article 25 of the Disaster Act.

The compensation to *businesses* implemented by the Flood Foundation Corporations Netherlands, a successor of the Floods Foundation Corporations Limburg. It negotiated with the ministry of Economic Affairs about settlements for businesses. In essence, the 1993 settlements were also applied in 1995, e.g., 65% of the damage to assets was reimbursed and a special credit settlement was set up. Furthermore, every corporation was considered for the reimbursement of DFL 1500 for the costs of evacuation. In case these costs were higher, more compensation would be given.

The communication between the ministry, the foundation and the corporations affected by the flood did not always run smoothly. The negotiations between the ministry and the foundation were disturbed by the statements of the Gelderland Chamber of Commerce. Via the media representatives from the Chamber stated that the economic costs of the flood for Gelderland were about DFL 700 million. These statements were not appreciated by the ministry and delayed the negotiations.

On the other hand, some corporations affected by the flood were not satisfied with the outcomes of the foundation's negotiations with the ministry. They claimed that their compensation was minute in relation to their real costs. Furthermore, the procedure to receive the compensation evolved slowly and was too bureaucratic. Dissatisfaction about the settlements was expressed especially by corporations from Gelderland. About fifty of them formed a special action group with the intention of achieving a better compensation deal, trying to hold the government responsible for the costs of the evacuation.

The ministry of Agriculture faced tough negotiations after the flood when it turned out that the *farmers* protested heavily against its compensation proposals. Demonstrations were held and heavy pressure was put on the minister to provide a better settlement for the farmers. Settlements were changed several times under the pressure of the farmers and their organizations. Eventually, nearly all the costs of evacuation were compensated. In addition, 65% to 90% of the damage to assets was compensated as well as extra costs such as cleaning costs.

Public versus private compensation

The 1993 flood — After the 1993 flood, the insurance companies introduced the idea of an "insurance pool". Cooperation with the government was deemed essential. Insurance companies would contribute DFL 50 million to this fund. This contribution would be made if the government paid DFL 200 million and provided fiscal facilities. The ministers of Finance and the Interior suggested that they would pay only DFL 80 million as well as offer some fiscal facilities. Additionally, they assured the insurance companies that the government would act as guarantee in case the damages would be above the total amount of the pool.[44] The insurance companies reacted negatively. They perceived the contribution from the government as insufficient, arguing that without sufficient resources no satisfactory insurance policy was possible. Negotiations were discontinued.

The 1995 flood — Following the 1995 flood, the government reopened negotiations with the insurance companies. Several members of the Parliament expressed their dissatisfaction with the lack of insurance against natural risks. Different options were studied using experiences from other countries. An agreement was eventually reached between the government and the insurance companies about the organization of a disaster fund. It was agreed to provide DFL 500 million in coming years. Every private person and corporation in possession of an insurance policy against damage for buildings and furniture could turn to this fund in case of damage caused by: earthquakes (more than 3.5 on the Richter scale) and fresh water floodings (bursting of or running over of dikes or a water level of 45,5m at Borgharen in case of the nonembanked part of the Meuse). The contributions to this fund would be paid by an increase in the premium on fire insurance policies. The maximum amount of financial compensation could not exceed DFL 500 million per annum.

In the end, a negative advice by the State Council aborted the agreement between the government and the insurance companies. The Council ruled that it was not appropriate to let all citizens pay for risks that only a part of the population faced. This implied that the government would have to create a new arrangement for the compensation of damage after a disaster. Preparations for such a arrangement were made at the time of writing this study.

Litigation and compensation: concluding remarks

It was hard to determine the total costs of the floods. Public authorities were willing to compensate for the damage but needed exact damage assessments, which were not available. Property damage as a result of water was visible and therefore relatively easy to assess (direct costs). In contrast, establishing the expenses of shutting down a factory or evacuating business property (indirect costs) remained difficult. There was a continuous discussion about the scope of these costs. Especially in 1995, protests continued for a better compensation of indirect costs due to the evacuation. Every victim group had its administrative counterpart: citizens (ministry of the Interior), corporations (ministry of Economic Affairs) and the farmers (Ministry of Agriculture). Since every ministry was familiar with the special needs of its clients, specific settlements could be set up. The ministry of Agriculture knew more about the specific demands of the farmers than any other organization. However, there were some differences in the outcome of the settlements. Every organization used its own definitions of "evacuation costs" or "cleaning costs". As a result there were differences in compensation given to the victims. This became even more pressing when it appeared that the National Disaster Fund used its own criteria for compensation. Parliament could not influence the rules by which this fund allocated compensation. The National Disaster Fund pursued its own policies, and various members of parliament threatened to limit the role of the Fund in future events, if the Fund would not arrange an extra settlement for evacuation costs.

In a strict legal sense, there was no law that made the government liable for damages caused by natural disasters. Hence, compensation of damage in the Netherlands is dependent upon the liberal attitude of governmental agencies (e.g. National Disaster Fund). Ultimately, the 1993 and 1995 damage compensation depended upon generosity, not on any consistent policy or legal liability. Generosity was enforced by political pressure, however. The government was perceived by many to have neglected the maintenance of the river dikes. Being generous to flood victims was therefore not only the decent but also the politically opportune thing to do.

Notes

1 Data for this report have been gathered using a combination of strategies. First of all, members of the Crisis Research Center have observed the decision making process fist-handin various provincial, regional and municipal crisis centres during both floods by sitting in on key meetings and taking detailed notes. Second, 75 extensive follow-up interviews were conducted with key decision makers at various levels of government and in a variety of counter-disaster agencies. Third, a comprehensive collection of public and confidential records of the crisis centres, evaluation reports, minutes of local council debates, other investigation reports and media coverage has been compiled and content-analysed. Fourth, both in 1993 and 1995 a phone survey was conducted among representative samples of the population in the various stricken areas. Furthermore, for the damage assessment and liability part of the study interviews were conducted with business representatives in all affected areas following both floods.

2 The villages of Borgharen and Itteren are parts of the municipality of Maastricht.

3 'Politiek Ramptoerisme', *De Limburger*, Maastricht, 29 December 1993.

4 The (approximately 110) polder boards are obliged by written constitution and the 'Polder board Act' to attend to the maintenance of dikes, roads, bridges and the navigability of canals. Like provinces and municipalities, polder boards are public authorities. The polder boards are independent and have their own competence domain. Moreover, they can raise taxes and issue binding rules pertaining to this domain.

5 During high water periods, the polder boards assess the safety level of the dikes and estimate the possibility of a breach of a dike.

6 Rosenthal, U., et al., *Complexity in Urban Crisis Management: Amsterdam's Response to the Bijlmer Air Disaster,* London 1994: James & James, p. 35.

7 Commissie Boertien, Deelrapport 3 (deel 2), December 1994, p. 2-1.

8 Commissie Boertien, Deelrapport 3 (deel 2), December 1994, p. 2-1.

9 The Institute for Inland Water Management and Waste Water Treatment (RIZA) is the research and advisory institute of the RWS for fresh water in the Netherlands and a national knowledge centre for integrated water management.

10 *Verslag Hoogwater 1995*, Evaluatie van de berichtgeving door RIZA, p. 2 (Evaluation Report).

11 'Rijkswaterstaat pleit voor kortere lijnen met buitenland', *Alert* nr. 2, February 1994, p. 6.

12 'Een Ramp in de Herkansing', *De Limburger Maastricht*, 31 December 1993.

13 Commissie Boertien, Deelrapport 3 (deel 2), December 1994, pp. 4-2.

14 Commissie Boertien, Deelrapport 3 (deel 2), December 1994, Bijlage 2B.

15 'Een Ramp in de Herkansing', *De Limburger Maastricht*, 31 December 1993.

16 'Interview with J.C. Van Haperen, head of the section public relations of the Regional Coordination Centre', *Alert* nr. 2, February, p. 12.

17 The risk assessment of the so-called 'unofficial predictions' is based on measurements with more unpredictable variables, reducing the reliability of the predictions. Considering the fact that these forecasts are less reliable, RWS only offers them under embargo.

18 Among others in: R.W. Perry, *Comprehensive emergency management: Evacuating threatened populations*, Greenwich: JAI Press 1985, p. 42.

19 This survey was conducted among respondents of three (Fire) regions: South-Limburg, Middle-Limburg and North-Limburg. We interviewed 50 persons in each of the regions. The sample was pooled: we only approached respondents in residential areas of the municipalities which were hit by the floods. After the selection was made under the residential areas, we took a random sample by means of phone books. Approximately half of the respondents were between 35 and 65 years-old. About a quarter of the respondents was between 18 and 35 years-old. The group of respondents older than 65 years accounted for a quarter of the respondents as

well. In this last group one out of three was older than 71. This implies that the group of evacuees interviewed is fundamentally similar to the average household in Limburg.

20 An a-select sample of the entire population in the stricken provinces were interviewed by phone. In Limburg, 420 persons who had actually experienced some kind of disruption from the floods were interviewed, 230 in the North, 60 in the Central region, and 130 in Southern Limburg. In Gelderland, a similar sampling strategy was followed for inhabitants of the Rivierenland (401 persons) and Nijmegen regions (104 persons).

21 Ministry of Transport and Public Works, *Onderzoek watersnood Maas*, Deelrapport 2, December 1994.

22 Commissie Boertien, Deelrapport 3 (deel 2), December 1994, p. 3-1/3-2.

23 See further, P 't Hart, U. Rosenthal, and A. Kouzmin, 'Crisis decision making: the centralization thesis revisited', *Administration and Society*, 25, 1993, 12-45. See also chapter 1.

24 *Evaluatierapport Nijmegen*, p. 21 (Evaluation Report).

25 *Evaluatie gebeurtenissen watersnood 1993*, gemeente Arcen & Velden, p. 28 (Evaluation Report).

26 Commissie Boertien, Deelrapport 3 (deel 2), December 1994, pp. 5-16.

27 *Evaluatierapport Nijmegen*, p. 33 (Evaluation Report); *Evaluatierapport Rivierenland*, p. 27 (Evaluation Report).

28 *Evaluatierapport Nijmegen*, p. 53 (Evaluation Report).

29 *Evaluatierapport Rivierenland*, p. 29 (Evaluation Report).

30 Letter to the Second Chamber, Department of Home Affairs, *Evaluatie Wateroverlast*, 6 October 1995.

31 *Evaluatierapport Nijmegen*, p. 53/55/56 (Evaluation Report); *Evaluatierapport Rivierenland*, p. 28/29 (Evaluation Report); Interview 19, 25 and 31.

32 Mr. ing. K.P. Barnhoorn, Overstroming en aardbeving (deel 1), in: *Beursbengel* nr. 637, August 1995, p. 20.

33 Commissie Boertien, December 1994.

34 Report of the Foundation Floods Corporations Limburg, 1994 and Commission Boertien, December 1994.

35 Rapportage financiën wateroverlast 1995.

36 *Algemeen Dagblad*, 27 December 1993.

37 This organization is a private initiative and deals with distributing indemnities after a disaster. The fund depends upon gifts from private persons and government. This fund does not apply for groups like corporations or farmers, but is solely meant for citizens.

38 Letter to the Second Chamber, Department of Home Affairs, 27 January 1994.

39 Letter to the parliament, Department of Home Affairs, 27-1-1994; Commissie Boertien, December 1994.

40 Vergaderstukken Tweede Kamer, 23 564, nr. 3.

41 *Ondernemen na de watersnood*, p. 10.

42 Commissie Boertien, December 1994.

43 *Financieele Dagblad*, 28 and 30 January 1995.

44 The government acts as a reinsurer. One of the main problems with natural risks for insurance companies is that, so far, finding a reinsurer for natural risks has proven impossible. The extent of damages after natural disasters has risen exponentially from an in international point of view. Reinsurers are therefore reluctant to provide reinsurance in these cases.

Chapter 5

FLOOD MANAGEMENT IN GERMANY

Wolf R. Dombrowsky and Lutz Ohlendieck

1 Introduction

This chapter describes and analyses the riverine floods in Germany in both 1993 and 1995. The analysis includes a comparison of disaster management during both floods. The floods affected the majority of the German states (12 out of 16 states), six of which are situated just within the catchment area of the Rhine and its tributaries.

In Germany disaster management is the responsibility of the states. The Federal Republic of Germany consists of 16 federal states. Within these states, the Ministries of Interior, as top ranking agencies (the so-called *Oberste Katastrophenschutzbehörde)* execute this task. In practice disaster management is delegated to local administrations at the communal level within the respective federal states (the *Unterste Katastrophenschutzbehörde).*

The report concentrates on two areas which were affected the most by the riverine floods: the city of Koblenz in the state of Rhineland-Palatinate (Rheinland-Pfalz) and the city of Cologne in the state of North Rhine-Westphalia (Nordrhein-Westfalen). Both municipalities are situated along the river Rhine and thus were directly threatened by the riverine floods. During the flood events of 1993 and 1995 none of the affected states, including the two cities mentioned above, formally declared the state of disaster, although the media, the public and even officials frequently used the term "disaster" to describe the unfolding events. The reasons why both cities did not declare the state of disaster ("state of emergency") although (large) portions of the cities were flooded in 1993 and 1995, will be explained in detail in paragraph 5.

Flood response in Cologne and Koblenz is analysed for both the 1993 and 1995 floods. The analysis will focus on learning from experience within municipal systems of disaster management (intra-system learning) and learning from other municipalities (inter-system learning). The two cities of Cologne and Koblenz differ considerably in terms of their disaster response system. Nevertheless, both cities surprised in 1993 by unusually fast rising floods and experienced similar response problems. Improvements and changes which had been made in the aftermath of the 1993 flood, allowed for a swifter response in the

1995 floods. The disaster management structure and organization were more decentralized in 1995. Accordingly more responsibility was passed to the regional or local operation centres and more competencies were granted to them.

The systematic gathering of data on the floods is impeded by the nature of German federalism and the subsequent fragmentation of the disaster response system. This lack of systematic data is largely due to the fact that disaster management is executed on the communal level by relatively autonomous organizations on a predominantly voluntary and honourary basis. Therefore, unified, centralized data bases, aggregated operation reports, protocols and records, and/or nationwide data on discharges and water gauges are difficult to access or not available at all. The differences with regard to countries with centralized organizational structures or specialized institutions (like the polder boards in the Netherlands) in this respect are obvious.[1]

2 The 1993 and 1995 riverine floods: an overview of events

The Rhine and its tributaries

The Rhine is one of the major rivers of Europe. With a total length of about 1,320 km it serves as an artery for water supply and transport, for sewer and drainage and for settlement and urbanisation. The catchment area of the Rhine, an extensive river system with its tributaries, is quite heterogeneous with respect to geologic, geographic and meteorological conditions. The area covers approximately 185,000 km² and stretches about 700 km from the Alps to the North Sea, and covers parts of Switzerland, Liechtenstein, Austria, Germany, France and the Netherlands.

The severe floods of 1993 and 1995 foremost hit the areas of Koblenz at the confluence of the Moselle and the Rhine, and Cologne downstream the incoming Sieg. The meltwater or drain from the Alpine Rhine did not cause the 1993 or the 1995 floods. Both floods were fed by the drain of the middle range mountains along the Upper and Middle Rhine.

Outline of the floods

The 1993 floods — From Constance to the area around Mainz, a complex system of dikes with different heights stands along the embankments of the Rhine. These dikes cover more than 500 km of the river's course. The probability of floods is estimated at one flood in every thousand years in the upper region, and every 200 years in the lower region.

To compensate for the different heights of the barrages, additional retention areas, weirs and floodable polders have been created. In many places, tall winter dikes are situated in more remote areas. Close to, or right on the embankments along the Rhine river between Lake Constance and Emmerich (near the German-Dutch border) only low summer dikes have been erected. These dikes protect against the relatively low summer floods and are not intended to withstand higher winter floods. Such winter floods are drained away to the retention areas. This is done on purpose to reduce flood peaks.

In the area downstream of Mainz to Bingen, the dikes protect against floods (probability of one flood in every hundred years). Downstream of Bingen and around Cologne there is only a very limited stretch of dikes. Cologne itself is secured by 16 km of river dikes and 11.4 km of flood protection walls, 1.4 km of which is mobile to maintain the scenic, touristic view of the Cologne waterfront. The mobile flood protection wall on the embankment of the Rhine is placed along and around the old town centre. These protective measures are only taken temporarily when water levels rise above 9 meter. This protection will be extended by the mobile flood protection wall to a height of 10m. Downstream Cologne, a complex system of summer and winter dikes and retention areas exists, offering protection against floods with a probability of one occurrence every 500 years.

In December 1993, the monthly precipitation in many areas in Germany reached levels of 200% above the mean based on a thirty year average (1951-1980). Several German weather stations recorded even above 300% to almost 400% precipitation above the mean. These weather conditions resulted in severe floodings along the Rhine river, mostly downstream. The flooding of polders enclosed by summer dikes upstream brought some relief there. However, the discharge of the Nahe and the Moselle into the Rhine further downstream caused the waterlevel to rise very fast to unusual heights. When the flood hit Cologne, it reached above the protection wall (10.63m). The river inundated the old town centre for almost three days.

A few kilometres downstream from Cologne, the comprehensive system of winter dikes prevented a wide stretch of terrain from flooding and severe damage. None of the dikes in the Rhine area collapsed or was seriously damaged. A few dikes were flooded or softened and undermined by seepage. Many of these dikes were originally built after the high flood in 1926 (10.69m in Cologne) or soon after World War II and were in need of repairment. For the tributaries of the Rhine River, no large scale flood control system exists. There are only singular dikes upstream, especially along the major adjoining towns. Most of the tributaries only have dikes where they flow into the Rhine. People living on the embankments or near the tributaries of the Rhine have developed their own strategies to cope with the almost annual floodings.

Apart from the municipality of Cologne, Koblenz was also hard hit, experiencing the worst flood since 1925/26. About 4,000 houses and almost 10,000 inhabitants of Koblenz were directly exposed to the flood, while approximately another 9,000 inhabitants suffered damage because their basements were inundated. The municipality of Koblenz is surrounded by middle range mountains which descend into both the Rhine and the Moselle. The city of Koblenz on the confluence is connected to several surrounding villages (town districts) which are incorporated in the municipality of Koblenz. The main city and the small villages are linked by roads on the embankments of the river. At certain water levels, however, the roads will be flooded, disconnecting these villages from Koblenz and each other. They became small islands surrounded by water and mountains. In Koblenz, no system of dikes or flood protection walls exists.

During the flood in 1993, most of the stricken villages were isolated. This demanded a flexible and decentralized leadership within the organization of the disaster response in Koblenz. Fortunately the social structure within the respective isolated town districts stimulated self-help and neighbourly assistance. The volunteer fire brigades proved to be up to the test, allowing for an effective and autonomous flood management in those places.

The 1995 floods — The 'Christmas Flood' of 1993 was characterized by a period of ten days with intensive rainfall saturating the ground. The 1995 floods were different. January 1995 was frosty and temperatures stayed mostly below freezing point. There was little precipitation, mostly in form of snow, down to the lower areas. The hurricane 'Ornella' then brought a low pressure zone over Germany and caused heavy rainfalls during 9 and 10 January in the low-lying areas of northern and western Germany. In the mountain areas, the snow started melting due to rising temperatures. The thaw resulted in extensive saturation of the ground. Additionally in many areas the ground was still frozen. The thaw and the rain, as well as the frozen grounds, in combination effectively sealed the ground.

On 21 January, heavy rain- and snowfalls infiltrated from the west. These showers came down on saturated ground, leading to an immediate surface drain to the rivers. The storm 'Thalia' brought a further low pressure zone over Germany and pushed the polar front further south. Masses of humid air passed along the polar front and caused heavy rains over the northern and middle part of Germany. Peaks of more than 50 mm of rain were measured on 22 and 25 January in Rhineland-Palatinate. The heavy rains lasted for days and covered a huge area, including most of the catchment area of the Rhine.

The strong precipitation combined with the thaw caused a rapid rise of the water levels in the rivers, amounting to more than 5 meters in Trier in 24 hours.

In Cologne, a "record level" of 10.69m was recorded, 6 cm above the maximum level of 1993.

Nevertheless, the floods of 1995 were not as severe as in 1993. In most places along the Rhine and its tributaries the damage was less. Earlier and better preparation served to reduce the harmful impact of the floods. Moreover, the Moselle carried much less water than in 1993, while the flood waves from Rhine and Moselle did not combine to amplify. This left more time to the authorities and the population to organize the response than in 1993. In Cologne the movable flood barrier and the inundation of the old town centre were once again flooded. Better preparation and timely evacuation kept the costs of damages caused by the flood limited to just half of the 1993 amount.

The two floods compared

Both the 1993 and 1995 floods were triggered by natural conditions. Both floods caused considerable damage, predominantly economic losses. While the people and the authorities were taken mostly by surprise during the flood events in 1993, they were better prepared when the 1995 flood struck. The public discourse on the causes of the flood differed in the 1993 and the 1995 period. In the 1995 flood period, the attention was more focused on the increasing influence of human activities on the incidence of flooding. A interpretative shift took place, away from natural and toward man-made causes. This shift has been the subject of scientific and bureaucratic controversy. The Federal Institute for Hydrology argues that meteorological and hydrological factors were primarily responsible for the floodings. The Federal Department for Environmental Affairs on the other hand claims that man-made sealing of the ground and hydraulic engineering were major contributors to the floods. A similar argument is made in a report of a parliamentary investigation committee in the state of Rhineland-Palatinate. The media, environmental protection organizations, and a growing part of the general public also took the 1995 floods as evidence for the argument that water management and disaster prevention policies had failed.

3 Water management and disaster response in Germany

The organization of water management

The federal law of waters, the *Wasserhaushaltsgesetz* (WHG, 1976) comprises all legal matters in Germany concerning water conservation, water supply, transportation and navigation on waterways. The WHG aims for a nationwide water management with emphasis on the purity of ground and surface-waters, regulating property rights for the use of water, water reservations (nature reserves) and the use of waterways as means of transportation. The federal WHG is implemented by the states *(Bundesländer)* which may also devise additional water laws. Specific needs for water conservation and water supply are administered through special acts and administrative provisions.

Generally, the major waterways in Germany are subjected to federal control. The federal water and navigation administration *(Wasser- und Schiffahrtsverwaltung des Bundes*, WSV) consists of several agencies subordinate to the Federal Ministry of Transportation. By law it is their task to provide for the planning, construction and maintenance of the federal waterways and the federal owned ports and dams, and to keep the waterways navigable. Regional water and shipping administrations should execute and supervise these activities. The federal authorities are solely responsible for the navigability of the federal waterways and the rivers, such as the Rhine. The states are responsible for surveillance and maintenance of the embankments, dikes and retention areas; for the dissemination of hydrological information (expected waterlevels) to the public; and the issuance of disaster warnings. Within the states, the administrative areas, the districts, municipalities and communities each take their share of these tasks.

Except for the federal waterways which are federal property, all other waters like surface, coastal and ground waters are the responsibility of the states. The surface waters are divided into waters of first, second and third categories. According to the state water laws, waters of the first category are the property of the state, whereas waters of the second and third category are the property of the owners of the real-estate on the waters' embankment.[2] Although the water rights of the owners (i.e. private businesses and households) are curtailed by law, it is difficult for states to regain real-estate in for instance ecologically sensitive areas to improve water and flood management.

State and federal government run the management of water affairs by means of broad skeleton plans. The design of these plans involves the Ministries for Commerce and Transportation, the Ministries for Forestry and Agriculture, the Ministries for Environmental Affairs as well as their subordinate agencies. General plans or special plans involve different authorities, institutions and organizations on the state and communal level. These concern the subordinate administrations of the aforementioned ministries and the communal offices for water management.

The responsibilities and the supervision regarding water affairs and flood warning are delegated to the different state ministries. In general, the Ministry of Environmental Affairs is responsible for supplying the necessary hydrological information. The Ministry of Forestry and Agriculture is responsible for the construction of dikes, measures and precautionary measures in the river basin. This covers the maintenance of reservoirs, and increasingly the reconstruction of nature reserves, which can also be used as retention areas to reduce the peaks of waves during riverine floods.

Although the state-level ministries of the Interior are formally in charge of disaster management and the warning and public information, the defacto responsibility and execution is passed down to the communal level. Recovery and reconstruction policies are also the responsibility of the communes. Only in case of exceptional damage, both state and federal ministries of Finance and Economic Affairs may intervene by allowing remission of taxes to people who suffered damage as a result of floods (or other disasters) or by means of direct financial compensations and low-interest loans. Such state subsidiary funds are an important contribution to recovery, because no public or private insurance against natural risks exists in Germany, except for the state of Baden-Württemberg.

The Federal Institute of Hydrology *(Bundesanstalt für Gewässerkunde*, BfG) in Koblenz is responsible for scientific research in the field of water management, the development of forecast-models for waterlevels, and forecasts itself. It cooperates with other hydrological and meteorological institutes in Germany and neighbouring countries. The German Weather Service in Offenbach in Maintz *(Deutscher Wetterdienst*, DWD), which resides under the Federal Ministry of Transport provides the main meteorological information and scientific research.[3]

The major problems facing water management authorities and institutions in Germany are similar to those experienced by other industrial nations. Firstly, there is concern about the deterioration of the water supply and water quality. The increased use of ground in densely populated areas seals the ground, diminishing the area for waterinfiltration and reducing the capacity of the ground to hold moisture. The result is a percolation of larger portions of the water into the so-called zone of saturation (ground water) and the overall lowering of the watertable. This development causes a faster transport of rain, melted snow and ice into rivers and other surface waters, and is considered a major factor in the worsening of riverine floods. Secondly, the increasing amount of sewage and waste water cannot always be processed by the industrial and municipal sewer disposal and purification plants. This imposes a burden on the quality of surface waters, triggering pollution increases. Moreover, riverine flood by itself leads to considerable pollution when sewer systems overflow and sewage spills into the surface waters. Finally, the scarcity of available space in the river valleys has resulted in increasing loss of retention areas and vegetation, especially around the embankments along the major rivers. For the past decades, communities frequently destined former water retention areas as development areas for business, housing estates and single homes. This has contributed to the sealing of the ground (with the aforementioned detrimental effects), and it has diminished the potential of water retention areas to reduce the effects of riverine floods.

The organization of disaster management

Under the German Basic Law, disaster response is the duty of the states. The districts, administrative areas and municipalities are obliged by state laws to make provisions in case of mass emergencies. In principle, disaster response is a state competency. However, in wartime disaster response becomes part of the civil defence system. Civil defence, contrary to peacetime disaster response, is organized at the federal level and administered by federal law. The Civil Defence Act of 1976 integrates the disaster response system into the civil defence system. To enable states to fulfil additional duties and tasks during wartime, the federal government expands the peacetime capacity of states' disaster response with so-called "supplementary and extending forces". The latter form an integral part of the peacetime system, although they are funded from federal resources. In practice, the mixture of peacetime and additional wartime forces causes some problems when it comes to estimating the identity and total size of disaster relief personnel.

All 16 German states have implemented or are in the process of implementing Disaster Response Acts. These acts are often combined with the existing Fire Defence Acts and comply with the federal law, the Extended Disaster Response Act. At the time of the riverine floods in 1993 and 1995 some states (particularly the so called "new states" in eastern Germany) were just in the process of making or passing new Disaster Response Acts. In Koblenz and Cologne, the State Disaster Response Acts of 1981 and 1977 were still in force during the flood events in 1993 and 1995. However, the context of disaster management has changed in two ways. Firstly, the obligatory federal structure for the organization of the Disaster Response exist no longer. Secondly, the post-Cold War abolition of federal supplementary and extending forces imposes financial burdens on the communities. With tighter budgets of cities and communities, the financing of the the lower Disaster Response Authorities becomes increasingly problematic.

In peacetime, disaster response is the primary responsibility of the local fire brigades, which are public services, and private organizations on contract with the district government (such as the Red Cross).[4] The fire brigades operate on a volunteer basis in communes with less than 100,000 inhabitants and on a professional basis in communes above 100,000 (mainly cities). Their disaster response task is added to their regular services. Expenses for relief work done during disaster management are refunded afterwards.

Depending on size of the communities, one or more emergency and rescue services are part of the Disaster Response Force. Although under normal circumstances some overlap and competition may exist between the various rescue organizations, they cooperate during mass emergencies or a disaster. The Disaster Response Force has a large potential of mostly honourary and voluntary serving people. The German Federation of Fire Services counts about 1.35 million members. Units of the Federal Institution for Technical Aid have about 65,000 members. Other medical rescue and care services comprise: the German Red Cross with roughly 4.5 million members, and a series of additional volunteer organizations with approximately 2.8 million members in total. Following the unification of Germany, a nationwide review and reorganization of all services and relief organizations has been initiated and is still in process.

The primary disaster response units and installations are under the control of the Administrative Coordinator, who is, depending on the particular type of state administrative organization, either the director of the district *(Oberkreisdirektor)*, the district president *(Landrat)*, the city manager *(Oberstadtdirektor)* or the mayor *(Oberbürgermeister)*. The Administrative Coordinator is responsible for disaster response planning, including the preparedness of disaster response units as well as the actual disaster operations. The local fire chief is responsible for on-site emergency management, being familiar with the local geography, infrastructure, population, dangers and resources. When a mass emergency exceeds local resources the regional fire chief takes over.

The decision to declare a disaster is made at the district level, notably the Administrative Coordinator. After a disaster declaration has been issued, the Administrative Coordinator takes command. The fire chief will establish a local operation centre, while the Administrative Coordinator will organize the coordination of the various organizations and services involved in a technical operation centre. Although in all German states this model for Disaster Response is followed, the names of the actual coordination and operation centres may differ among the states.

When the situation is sufficiently serious, the Administrative Coordinator will establish a local disaster coordination centre with a disaster management staff. It comprises representatives of communal agencies like (public) transportation, roads and construction, health, water supplies and sewage, gas and electricity, the rescue services, and national organizations such as telecommunication, railroad and Army. The Administrative Coordinator is also supported by experts in fields of fire protection, recovery, maintenance and repair, medical service, veterinary matters, NBC detection and decontamination, care and control, communications, sheltering, provisioning and public assistance.

According to site proximity, a disaster area can be subdivided into operational areas and sectors with their own operation centres to add flexibility to disaster operations. These centres are supervised by the coordinator's disaster management staff. The coordination staff generally assumes the political and administrative responsibility for emergency management, including decision making and overall coordination, whereas the technical operations centre is responsible for the tactical leadership and coordination of the actual Disaster Response Units on the scene.[5]

Once the local emergency management resources supplied by the fire brigades and rescue services of a community have been exhausted, disaster relief is undertaken in the first instance on the district level (administrative area or municipality). The formal decision to declare the state of disaster is most likely when the scale of emergency exceeds existing local resources and outside assistance and coordination is needed. In such a case it is mandatory for neighbouring communities to give aid and assistance to their stricken counter-parts. The state will only designate the stricken region a disaster area and form a disaster management staff on the state level if several of its districts are struck by disaster and if their combined resources are not sufficient to cope with the emergency.

The system of disaster relief is basically executed by the regional authorities, which comprise the districts *(Kreise)*, the municipalities *(kreisfreie Städte)*, and the administrative areas *(Regierungsbezirke)*. Only if several regions are simultaneously struck by a disaster, state government will intervene and establish a state-level disaster management centre.

4 Risk communication

Intergovernmental warning

In all states, the flood reporting or warning service monitors and reports flood levels, and compares them with regard to pre-defined risk levels (so-called 'reporting-levels'). On the federal level, the Federal Water and Navigation Authority (WSV) maintains the service for flood reporting for federal waterways. All services gather and coordinate data for the protection against flood hazards and floating ice. Public information, including forecasts and warnings if necessary, is delivered by these services.

Each state has its own provision for flood reporting. Although the flood reporting procedures are not standardized, all states follow similar procedures for cooperation with the WSV. The WSV is obliged by the Federal Law of Federal Waterways to maintain a flood reporting service in cooperation with the states.

The water level and accompanying hazards are distinguished in different reporting and alert stages.

In the Rhine area and its wide catchment area, the WSV and six state services produce and deliver flood forecasts and warning procedures. The warning process in the states of Bayern, Baden-Württemberg, Hessen, Rheinland-Pfalz and Nordrhein -Westphalen is regulated by the state. In the Saarland the WSV is responsible for the flood reporting and warning. The Federal Institute for Hydrology, acting as the supervisory authority under the control of the Federal Ministry for Transport, is responsible for the preparation of models for the water level and water drain forecasts.

In the Rhine area, there are seven water level forecast centres situated in Würzburg (Bavaria), Karlsruhe (Baden-Württemberg), Saarbruecken (Saarland), Trier, Koblenz and Mainz (all Rhineland-Palatinate), and Duisburg (North Rhine-Westphalia). These forecast centres gather the information directly from the numerous water level stations by means of computerized data transfer via modem and telephone. The number of water level stations varies considerably among the states with approximately 300 stations in Baden-Württemberg and 38 in the Saarland.[6]

Almost all the WSV water level stations measure or estimate river discharges. However, not all these stations are fully equipped for the flood information service (between 8% of the situations in Rhineland-Palatinate and 52% in Hesse). In practice the water levels have proven to be more adequate indicators for flood development. In both Cologne and Koblenz, the local water gauge is the locus of measurement in the alarm and action plan of the Disaster Response. In order to estimate the development of local water levels, upstream water gauges of the Rhine and Moselle as well as its tributaries are watched. This allows for reliable forecasts up to 24 hours. Precise forecasts can be given only for six hours.

There are different forecast models in use for the different rivers to predict the water- levels and discharges of these rivers. One hour after relevant measurements are made, the forecasts for 6, 12, 18, 24, 30 and 36 hours are available through telephone and/or data transfer by modem. The reliability of forecasting syste-matically decreases as the length of the forecasted period increases.

Different ways of reporting and disseminating flood related information and warnings can be found, depending upon the legal, administrative and organizational structure of the states. The states of Bayern, Baden-Württemberg and Hessen differentiate their services into a central regional and a local service following the administrative structure of the water authorities. The local offices of the water authorities gather the reports of the different local water levels and pass them on to the different communities, districts, municipalities, governmental authorities and whoever may be concerned. In Baden-Württemberg, the police executes the task of the local water authorities; in Rheinland-Pfalz and partially in Hessen the police fulfils regional tasks. In general, the regional services inform the central services on the state level and those in the neighbouring states.

The structure of the flood information service and the warning to the public of Rhineland-Palatinate will be used as an example. In Rhineland-Palatinate, the flood information service was introduced in 1986 and is regulated by the Flood Reporting Provision from February 1986. This provision covers the rivers Rhine, Moselle, Saar, Lahn, Nahe, Glan, Sieg, Sauer and Our. All details for the execution of the Flood Reporting Provision are administered by the regional flood reporting plans. Rhineland-Palatinate has established three flood reporting centres to this end. Since federal waterways are involved as well (which, in the context of this study, comprise the rivers Rhein, Mosel, Saar and Lahn), state and federal authorities required by law to cooperate. The same goes for flood reporting arrangements with the neighbouring states of Hessen and Baden-Württemberg. With concern to the combined flood reporting the service observes the precipitation, water levels and discharges of the rivers. The data is evaluated and reported to the districts, municipalities and the public affected. This service delivers information and forecasts on the development and course of flood events to allow for timely local response and the swift taking of protective measures.

In Cologne, once the waterlevel in the rivers rises above a marked level at the water gauges, the flood reporting service is activated and a flood situation report is prepared. The subsequent "opening" report on the flood situation is immediately transferred to the districts concerned via the telecommunications of the Police of Rhineland-Palatinate and to the relevant municipalities via the Flood Reporting Centres. The district authorities are obliged to inform the communities and the public about the opening report and possible hazards. The administrations of the municipalities warn the public, businesses and municipal offices.

In practice, municipalities like Cologne and Koblenz receive information about upstream watergauges via Fax and/or BTX. City authorities make their own estimates apart from the available forecasts of the hydrological institutes. The most relevant water gauges for such estimates are local ones. The water levels at these water gauges has proven to be a better indicator for the flood development than the measurement of discharges. During 1993 and 1995 Cologne and Koblenz authorities mostly relied on their own forecasts.

The warning process during the floods

Up-to-date information on the water levels of the Rhine and its tributaries are available and made public by telephone service of the Telecom, via television, radio or electronic media. All information is drawn from the same sources to avoid or at least to minimize the chance of contradictory forecasts and confusion between various reporting services which subsequently could lead to confusion and feelings of insecurity of the general public.

1993 — The flood reporting service for the river Moselle was put into operation on 17 December 1993, for the Nahe and Glan on 19 December 1993 and for the Rhine on 20 December 1993. The services worked around the clock until the first flood wave subsided after Christmas. An official report and forecast on the flood situation, was made available to administrations and the public at all times via telephone service, Videotext, BTX and the local and regional newsradio.

The 1993 flood affected all districts and municipalities along the Moselle, along the Rhine downstream the district of Mainz-Bingen, the district Bad Kreuznach at the Nahe, the district Kusel at the Glan, and the town of Zweibrücken (through minor waters: the Hornbach and the Schwarzbach). Altogether 488 communities with 102,000 inhabitants were affected.

1995 — On Monday 23 January 1995, the flood reporting service for the Rhine River was established at 07.00 in cooperation with the state and federal services. Since the calculations based on the forecast models indicated that reporting levels at Koblenz (450 cm) and Maxau (650 cm) would be surpassed during the next 24 hours, the reporting service was established already before the reporting levels were reached. On the basis of weather forecasts and the course of floods in the tributaries, it was expected that the water level at Koblenz would exceed 700 cm within the next 24 hours.

The flood reporting service went into fulltime operation, starting on 25 January 1995, in response to the fast rising water levels and hazardous flood waves, and remained so until 1 February. During the flood, an additional flood report besides the regular daily report was prepared and disseminated every evening. Already on 25 January 1995 a warning was issued that the downstream levels of the Rhine at Koblenz could reach or even rise above those of the 'December Flood' of 1993. Altogether 22 flood reports were disseminated by the flood reporting services.

Informing the public

At times of crisis or disaster, accurate and reliable information is a key to successful disaster management. However, people do not only wish to be kept informed on general aspects of floods, they also want to receive detailed and specified information, with clear and practicable instructions, to be able to cope with flood hazards. The Ministry of Environment of Rhineland-Palatinate, as well as the municipalities of Cologne and Koblenz distributed flood instruction leaflets informing the citizens in detail about information sources (phone numbers, frequencies, addresses, codes etc.), access to these sources and measures to be taken in case of floods. It also gave information on protective self-help and how to equip a household in the case of a breakdown in public supply. Futhermore, it listed the major official sources of public information which are available in the city of Cologne, including telephone information services; regional and local FM radio stations; and water level information on TV via Videotext in three nationwide and regional programmes.

The municipalities of Koblenz and Cologne also put up posters with the above information in flood prone areas. During the 1993 flood, special issues with the latest information on the flood were put up on public notice boards in Koblenz. Compared to former floods, this did not prove to be effective due to the fast rising water levels and subsequent change of events. Often the news was already out of date prior to distribution. So this practice was not continued during the 1995 flood. Since in 1993, telephone communications were disrupted for many households in Koblenz, special Fire Service vehicles with loudspeakers drove through town announcing the latest news and issuing instructions. Additionally, boats patrolled every half hour in those areas where access by vehicles was impossible. Again the people received instructions and were asked to put out a cloth or flash lights at night to signal the patrolling boats for help in case of such things as illness, and shortage of food or other supplies.

The principal scheme of public warning in case of river-floods remained unchanged during the 1993 and 1995 floods. One substantial difference was the improvement of the infrastructure. In Koblenz the number of radio sets was doubled, the reserved telephone lines for the connection of the different operational centres were made waterproof by installing them above the flood level, and the number of telephone lines was increased. The 430,000 telephone calls to and over 30,000 BTX-on-line-contacts directly with the authorities, clearly show that qualitative aspects of information have increased in importance.

Informing the media

The collaboration between media and authorities regarding warning and informing the public was considered to be good. By law the media has to transmit prepared messages, warnings and information according to standard procedures.

The relationship between government and the media has been overshadowed by controversial interpretations of the causes of the floods and conflicting views of the news coverage. Tendencies to sensationalise and dramatize the flood events in the way of "action news" and "reality-TV" provoked harsh criticism. Nevertheless, the cooperation of the printed media, the local radio and local TV with the warning services and the Disaster Response Officials was considered good. The national news took a more sensationalist "infotainment" line.

A small group of ruthless disaster tourists actually tried to sabotage flood protection devices. A large majority of the estimated 6.000 disaster tourists in Cologne were harmless onlookers. Nevertheless, they were not allowed near the mobile flood protection wall any more, after some of them had tempered with the structure.

During the floods daily news coverage by the press was focused on local events. It acknowledged the work of the Disaster Response and the combined effort of the people, victims and helpers to cope with the situation. Simultaneously, nationwide TV-coverage of the flood events had a more decisive influence on the collective perception of the victims' tragedy and the failure of flood control. Many onsite reports included interviews with politicians, officials, victims and relief personnel. These were frequently followed by contrasting interviews with politicians and scientists, documentaries on water management, the greenhouse effect and global warming, and critical reports on political and administrative shortcomings.

Many TV-documentaries showed that, due to the scientific data and knowledge available on meteorological, hydrological and ecological developments on a global scale and their regional effects, there is strong evidence for the anthropological coresponsibility for the increasing frequency of riverine floods and especially flashfloods. Nevertheless, statements of politicians favour the argument that there is no clear evidence for man-made causes of riverine floods. They seem to rest with their opinion on the lack of clear statements on the side of the scientists. Scientists find it rather difficult to come up with solid proof, since they focus on the creation of new and more complex models for the understanding of the global climate and its regional effects.

On the other hand the media discovered ecological disasters as a favourite topic, and they rather tend to follow the argumentation of the more critical scientists and the environmental protection groups, who clearly define the man-made factors to be mostly responsible for the floods. As a result of the frequent and critical coverage of local and global events like civil wars, environmental pollution problems, acid rain, the greenhouse effect/global warming, water shortages, spoiled crops, etc. there is a growing public sensibility towards topics such as induced disasters, their man-made causes and their devastating potential in the near future.

In 1993 there was a massive breakdown of the communication of the Disaster Response in Koblenz due to insufficient radio equipment and the collapse of the telephone system. Telephone cables and distributers were installed underground or just above the ground, and they were not waterproof. The same applied for the power lines which were damaged and left the households in the flooded areas without electricity. As a result, private radios, TV-sets and PC's could not be used any more to gain information via the news, BTX, Modem or Fax. These shortcomings made the coordination of the disaster Response in Koblenz at times impossible, but after the 1993 flood these problems were solved and during the 1995 flood none of these mishaps occurred again.

5 Disaster response

General overview

Since disaster response is initially a communal responsibility, we will focus on the response of the cities of Cologne and Koblenz. Neither the states, nor the municipalities of Koblenz and Cologne formally declared the state of disaster in 1993 and 1995. However, the two cities did activate their disaster response forces in 1993 and 1995 and both cities received assistance from disaster response forces of neighbouring districts.

Despite severe damage to town centres and districts in both cities, no formal declaration was issued in 1993 and 1995. In Cologne the water level at the Cologne water gauge remained just below the disaster alarm threshold of 10.70m. Although a disaster was not formally declared, in practice the situation was dealt with as a disaster.

The Lower Disaster Response Authority of Koblenz did not declare the state of disaster since the State Disaster Response Act of Rhineland-Palatinate does not allow for such a declaration. This has been done to prevent organizational and communicational disruption when the leadership and disaster management structure is changed and shifts are made from the "normal" to the "disaster" administrative process. Hence, communities and districts in Rhineland-Palatinate are provided from the outset with additional legal powers for disaster management, such as the authority to issue instructions to the communal offices and services.

The Christmas Flood of 1993, with its unexpectedly fast and high rising water levels took many people, authorities and scientists by surprise. The rapidity and the force of the flood events hit an inadequately prepared and protected population. In many places, the protective measures, the disaster response units and volunteers and the material and the infrastructure were insufficient to meet the demands of the population and to secure the continuous supply of the people in all stricken areas at all times. Although few lives were lost, severe damage to private and public property was done.

Soldiers of the German Bundeswehr remained stand-by in the affected areas from 22 January until 6 February 1995. The army fulfilled important support functions such as preparing sandbags, constructing temporary dikes, rescuing and evacuating people, operating a ferry service with rubber dinghies and ferries, supplying civilians in isolated spots, and securing and clearing affected areas. The cooperation between the armed forces and the disaster response units, i.e. the coordination through the disaster and tactical coordination centres was established immediately and worked well. This not only holds true for the German Forces but also for the 600 French troops who were in action for six days and the 200 US-soldiers who were in action for two days. Their assistance was efficient and relieved the disaster response forces. Additionally, the German Air Force employed Tornado reconnaissance aircraft to take aerial photographs and map the extend of the floods of the Rhine and Moselle river. The Federal Ministry of Transport sent aircraft for aerial exploration which are normally used for the surveillance of maritime pollution and which are equipped with modern sensor technology to detect the spillage of oil. Moreover, the Federal Institute of Hydrology (BfG) measured the speed of flow and the discharges of the Rhine with ultrasound equipment.

Disaster management in Koblenz

Since the disaster law of Rhineland-Palatinate does not provide for a declaration of Disaster, the administrative coordinator is free to decide when to declare a "mass emergency" and to legally oblige the army to support the disaster response. The declaration of a mass emergency thus becomes a strictly intra-administrative and formal procedure. It is not needed to obtain extraordinary legal powers. The authorities in Koblenz nevertheless used the term 'state of disaster' unofficially, mainly to facilitate communication with the media and the public.

The 1993 response — The organization of management involves the type and division of the management levels, their coordination, as well as the distinct responsibilities of the different levels and heads of operation. During the 1993 floods, the administrative coordinator met with the central coordination staff twice a day. During the staff meetings, the staff consulted with the representatives of the relevant city offices, the rescue services, liaison personnel of the police and the Army, and the heads of the technical and the area operation centres who delivered a situation report. In the staff decisions were taken on political and administrative matters of disaster management, whilst the technical (operational) coordination centre implemented the staff decisions. The operational centre was situated in the same building as the staff. Subordinate to the technical coordination centre, three area operation centres managed the operations in sections of the city. In turn, these sections subdivided into local districts, each with its own local operations centres (see figure 5.1).

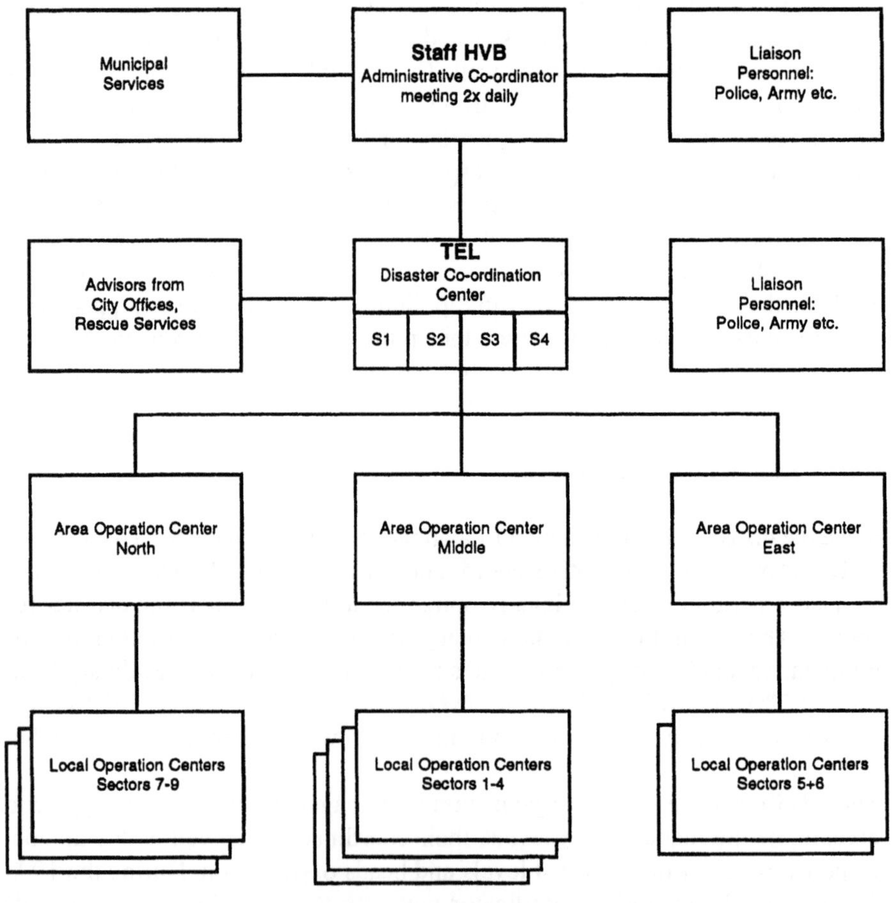

Figure 5.1 Flood response organization in the city of Koblenz

In 1993, these operation centres and the Area Operation Centres were tied more strictly to the central leadership of the Staff HVB/TEL. This caused severe problems, since the central command structure was interrupted at times due to a complete breakdown of the communication infrastructure. The area operation centres and field units were cut off from up-to-date information on the flood development and available resources.[7] Underground telephone lines became unusable due to flooding and rising groundwater. Moreover, there were not enough radio sets to compensate for this loss of communication lines. Telecom supplied the Disaster Response Force with 18 handsets. However, these proved to be of little value because their frequencies were also used by administrators and private households.

The planning for disaster response in Koblenz, like in Cologne and other places, was based on the experience of the past decades. Since 1948 the Rhine area had not been struck by extraordinary high and lasting floods. The floods of 1983 and 1988 remained significantly below the 9m water level in Koblenz, the "critical" level for the planning and execution of protective measures. Up to the Christmas Flood of 1993, the floods followed similar patterns, with slow rising water levels between 6m and 8m allowing at least two to three days preparation time, in most cases even a week. This slow rise of water allowed disaster response to work according to "flood reaction plans" attuned to different water levels.

The Christmas Flood of 1993, however, turned out to be different. The fast rise of water cut back preparation time to circa 20 hours during 22 December 1993. The old flood reaction plans were not designed for such a short response time. In addition, a large number of the disaster response personnel was on Christmas holidays, many soldiers were discharged before Christmas. This lack of personnel caused problems, the available equipment (gangplanks, sandbags and pumps) could not be timely installed on the scale needed. Moreover, the Area and Local Operation Centres for the different areas and sectors of Koblenz could not fully operate on time. Only 30% of the disaster response personnel was available in the first 24 hours. Support by the army was delayed too, since with reduced personnel the army had to secure its own material first. Additionally, there was a 40% increase of regular fire fighting and technical aid missions during the 1993 flood, compared to earlier floods, which drained away considerable capacity (the professional fire service) from the already diminished disaster response force.

In Koblenz, at the confluence of the Rhine and Moselle river, the water level reached its highest peak since 1924 on 23 December 1993. Large portions of the city were flooded. Many people in Koblenz were cut off from the outside world: there was no electricity, telephone lines were disrupted and there was a lack of supplies. Oil-spills into the flood water caused unpleasant vapours and polluted the ground soil and ground water, as well as public and private properties. The situation for the people affected and the relief forces deteriorated as temperatures dropped to around 3-4°C.

More than 25,000 inhabitants were affected by the flood, 10,000 inhabitants were temporary cut off from the outside world, and about 7,000 inhabitants were without electricity. Many of the victims who left their homes due to the cold after the failure of the heating systems were accommodated by neighbours or nearby family. In Koblenz only about 300 persons were evacuated. Overall, there was great solidarity among the people. The willingness to help was overwhelming.

The lack of sufficient capacity of the staff and response units impaired a sufficient supply of the population during the Christmas days. The local centres were located in public buildings, schools or stations of the fire brigades, and operated satisfactorily. It was therefore deemed unnecessary to operate the Disaster Coordination Centre for Koblenz continuously. Two meetings a day at 10.00 and 16.00 were deemed sufficient. At those meetings the heads of operation of the three flood management areas and the representatives of the different services and offices involved, presented their situation reports and offered their judgement on the situation and possible evacuations. Necessary measures to be taken were discussed, and requests for personnel, equipment, material and provisions were made for the forthcoming hours.

Tasks were delegated, so the tactical, the Area, and the Local Operation Centres could concentrate on the disaster management on-site, while the overall coordination centre focused on political and administrative problems and decision making. Aside from personnel shortages, the main problem besetting the operations was communications. There was a shortage of major telephone lines, six lines for each area control centre, and as explained before, this lack of telephone lines could not be compensated by mobile phones. The number of authorized radio frequencies (channels) for the disaster response forces proved to be too low. The lack of hardware in terms of radio sets and walkie-talkies; the lack of frequencies available; and the fact that reinforcements from other organizations or areas did not operate on the same frequencies and used preset fixed channels caused severe communication failures and hindered the coordination. The breakdown of more than 7,000 telephone lines due to flooding and the rise of the groundwater, forced the rescue services to rely strongly on radio communication, which was overburdened already. An additional problem was that the radio sets of the German Army, i.e. their available frequencies, did not match the frequencies of the other services. This made communication impossible until the Fire service lent several of its own radio sets to the army.

A further problem was media management. The spectacular events and the obvious response problems attracted massive interest of the media. Most rescue agencies were unprepared for this aspect of a disaster, with reporters and camera teams invading the disaster areas and competing for interviews with on-scene commanders. The rush took officials by surprise and many of them had to spent up to 70% of their valuable and limited time to answer questions to reporters, at the detriment of their primary duties.[8] Likewise, the work at the disaster response forces was severely impeded by disaster tourists. In this respect, the construction of elevated waterways for the inhabitants throughout the flooded parts of the town proved to be something of a mixed blessing, since it allowed an optimal access to the site for the disaster tourists. The obstruction of the disaster response units by onlookers and disaster tourists was not limited to dry land. Many of them took small boats clogging the flooded streets. Consequently, barriers had to be erected against this flottila of onlookers.

The 1995 response — After the flood events of 1993, the flood management structure remained basically unchanged. However, personnel received better and more intensive training to prepare them for fast and high rising riverine floods. Also, the communication infrastructure was improved substantially, while elements of cooperative and decentralized leadership were strengthened. The area and local operation centres obtained more autonomy to take tactical measures onsite. Decentralization avoids unnecessarily detailed communication with the superior levels, which inhibits a fast and adequate reaction to local problems. Another advantage is minimisation of potential misunderstandings, while decentralization also reduces potential information overload of the superior operation centres. In conclusion: disaster management response has moved away from a command and control system to a self-reliant and decentralized system which can adapt faster and more efficiently to the local problems and needs.

The means of communication were also improved substantially, as the numbers of telephone lines, radio sets and walkie-talkies, as well as authorized frequencies (channels) increased. By 1995 the telephone lines and distributors of the German Post/Telecom were made water-proof.

In 1995, a press office was established to relieve media pressure on operational commanders and other relief personnel. All people with questions concerning the flood events and disaster response were referred to this press office. This arrangement proved very efficient.

The constant flow of information between the water reporting stations and the operation centres was also improved. The alarm planning was adapted to the possibility of rapidly rising floods which necessitate earlier precautionary measures. This adaption carried political overtones since early precautions could turn out to be unnecessary afterwards, while having caused major costs and inconvenience. This was actually the case in 1995, when the water level remained below the level of 1993 and the city of Koblenz was not inundated.

The construction of walkways was delegated to private firms in order to reduce the workload of the disaster relief units. In 1995 the gang planks proved unnecessary due to the lower level of the flood wave. Some people complained about the discomfort caused by the construction of gang planks in the streets. However, the city authorities preferred to be on the safe side this time after the experience of the Christmas Flood of 1993.

The flood levels of 1995 rose slower and did not reach as high as in 1993, so there was enough time for preparations by the population and authorities. In January 1995 all personnel was mobilized and available, including the Army, so the disaster response units were more or less at full strength.

The problems with oil spillage were minor, since many households had switched to natural gas for heating and cooking. Moreover, existing laws for safety measures to be taken in flood prone areas, which included the securing of oil tanks, were tightened and enforced more strictly.

During the 1995 flood, organized forms of disaster tourism occurred for the first time as a major problem. Apart from the typical onlookers who impeded the mobility and work of the relief forces, especially in the often narrow streets of the old town parts, also organized forms of disaster tourism came to the fore. A private TV-station offered tours for DM 20 through the flooded parts of the town, including snorkling into the flooded homes and a T-shirt 'Ich war drinnen' ('I was in'). The authorities reacted harshly and threatened with severe fines, up to DM 10,000. Also the Rundfunk- and Fernsehrat, an official media ethics watchdog, was asked to take up the case. No further incidents of this kind took place afterwards. The TV-station claimed not to be responsible for the whole affair, since the broadcast was not produced by them but bought from a third party and was only meant to be a joke.

Volunteer organizations played an important role in Neuendorf, the most severely stricken part of Koblenz. Here an emergency action organization was founded within the neighbourhood which cooperated closely with the disaster response units. As a "convergent group" it agreed to be integrated into the command structure of the disaster response of the respective sector, thus becoming "disaster response helpers" with the status of volunteer firemen. For the volunteers, this also had the advantage of being insured. Their assistance was of great help and was a relief for the regular services in that sector, but so far this type of emergent action has been exceptional.

Disaster management in Cologne

No major floods had hit Cologne in the first 35 years after the second world war. This gave its inhabitants an illusion of safety and produced a certain complacency with regard to flood protection measures. The flood of 1983, rising to a level of 9.84m and flooding the old town centre, initiated public discussion about measures to be taken to protect the city of Cologne, especially the old town centre, without destroying the old townscape. The city government decided to raise the flood protection level to 10m. This decision was based on the 1926-estimate that only once-in-a-century floods would rise above the 10m level. To protect the old town centre, a mobile flood protection wall was built. This wall guaranteed protection up to the level of 10m, and proved successful during the 1988 flood which rose to a level of 9.95m.

In Cologne, the disaster response aims of the Flood Protection Centre and the Department of Sewage Treatment are intertwined. One of the tasks of the department of sewage treatment is to deal with water and to react as soon as possible when water levels start to rise. In addition, this department is responsible for the safe disposal of sewage during floods. During the flood of 1988, when the water level remained just below the protection level and inundation of the city was prevented, flooding of the sewer system resulted in tremendous spillage of untreated sewer into the Rhine river. The authorities remedied these problems by implementation of the plan *Abwasserkonzept 2000* (Sewer System 2000) which was already developed in 1987. The whole sewer system of Cologne was restructured and modernized at the cost of DM 650 millions to be able to cope with water levels up to 10m. Additionally a DM 200 million-draining system was installed and put into operation in 1993.

The disaster response to flood events in the city of Cologne is executed according to a basic frame for flood protection and response. This plan is attuned to the water levels measured at the Cologne Water Gauge (Kölner Pegel, KP). If water levels rise with an expected increase of more than 5 cm/h or if unusual precipitations and conditions may cause a rapid increase of water levels within a short period of time, the Department for Sewage Treatment is mobilized.

The Flood Protection Centre acts as a basic coordination unit for flood management in Cologne up to a level of 10.70m KP. The Centre is part of the Department for Sewer Treatment of the City of Cologne. Its permanent staff is made up of representatives, mostly administrators, of the municipal services which are responsible for flood management. These services comprise the Office for Sewer Treatment; the Transport Office (walkways, boat service, traffic guidance, securing dikes); the Construction Office (roads, tunnels, bridges, flood gates, underground transport); the Office for Housing (evacuation, shelter and provision); the Police Department (traffic guidance, securing flood protection measures and private property); the Fire Department (pumps, oilspills, coordination of rescue services, transport of the sick and for evacuation); and the liaison personnel of the German Defence Force (provision of population, transport and implementation of flood protection gear, boat and ferry service, walkways and pumps). Additional services can also be involved, depending on the development of the waterlevel, for instance the Office for Public Transportation, the Harbour Office, the Office for Gas and Electricity, the Office for Environmental Affairs, the Office for Health and Social Affairs, the Telecom (telecommunication) and the Press Office (compare figure 5.2).

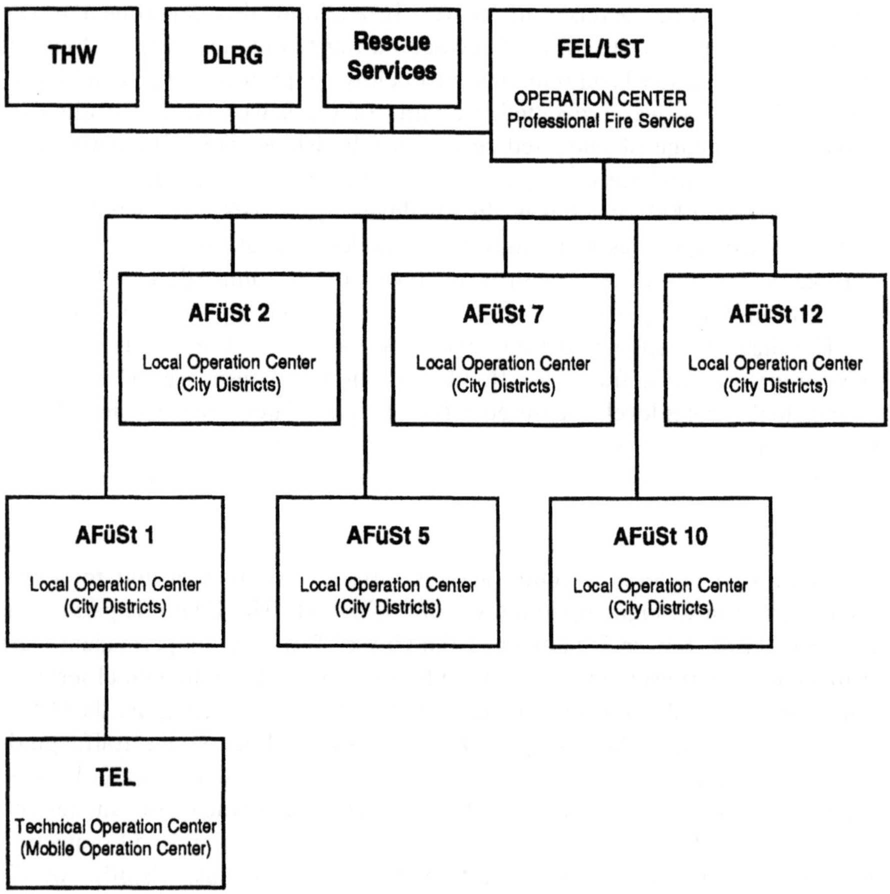

Figure 5.2 Flood response organization in the city of Cologne at moderate water levels

There exist a range of situations from "routine" flood management to disaster management, when the deployment of disaster response units is not yet necessary, but regular resources and structures for efficient management are insufficient. To cope with these mobilisation of staff and integration of other municipal services and offices into the Flood Management Staff is needed (see figure 5.3 next page). In practice, when the old town centre is flooded, the flood coordination centre is enlarged to a 'Flood Crisis Management Staff'. This is needed to handle the evacuation and provision of the population in the flooded town centre, but also to solve sewer and environmental problems caused by the flooding.

The Centre coordinates the flood management at the administrative level and is responsible for political decision making. Parallel to the Centre, the operational coordination of the response forces, i.e. the fire and rescue services, and the police is shaped by the operation centre of the Cologne fire service. The staff of the operational centre is enlarged when the water level reaches 8.30m to be able to increase flood protection activities, to secure buildings and public roads, tunnels, embankments etc. The expanded operational centre is similar to the Disaster Coordination Centre in Koblenz described above. It is connected with the overall coordination centre via reserved telephone lines and a fire service liaison officer to secure a continuous communication flow.

Six Local Operation Centres in the different town districts along the embankment of the Rhine operate subordinate to the operational coordination centre. These so-called Technical Operations Centres are accommodated in the local fire stations. They operate mostly autonomously and are responsible for the deployment and provision of their own personnel as well as the assigned personnel from outside Cologne. They communicate frequently with the fire service operational centre staff to update situation reports and to give information to the logistic coordinator for the restock of personnel, provisions and resources.

The 1993 response — The Flood Protection Centre at the department for sewer treatment started to operate on the morning of 13 December 1993. The Centre coordinated the information from the police, the fire service, and the other offices and organizations involved. It made forecasts on the development of the flood and passed this information on to the public, private firms and the media. The Centre also advised people who were threatened by the flood or who had suffered losses. Seven telephone lines were available plus one direct line to the fire service, and at peak times, i.e. at rising water- levels, up to 5500 phone calls were received in one day. Additionally the Centre coordinated measures that were taken by the department for sewer treatment at the different water levels.

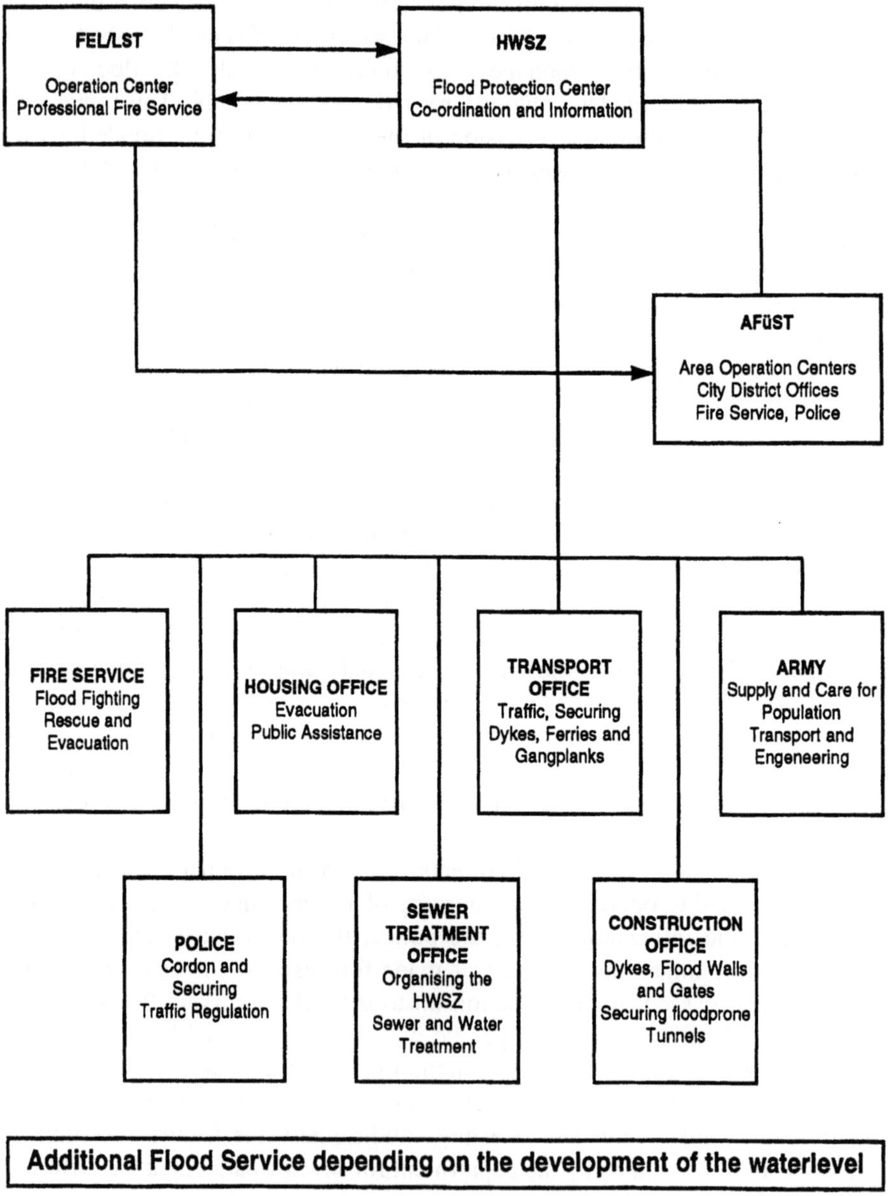

Figure 5.3 Flood response organization in the city of Cologne at high water levels

Telecommunication centres and control rooms were set up. All relevant municipal offices and services had day and night personnel on duty for flood protection and public counselling. For example, the municipal Gas, Electricity and Water service of Cologne which combines the power stations, gas- and waterworks received circa 3,000 telephone calls per day.

The different municipal offices and departments began their assigned flood tasks in the endangered parts of Cologne independently. Formal coordination started on 19 December 1993 when the water level reached 5.70m KP. When the water rose to 8.50m KP combined with negative forecasts, representatives of all offices and departments joined for a common staff meeting, to update the situation report and to coordinate the next phases of the operation. To reduce media pressure on the heads of operation and the response forces, a central press office was set up.

During the flood a large number of disaster response units from outside Cologne assisted local fire service. In the beginning problems occurred with the radio communication, since there was only a limited number of preset channels available for the radios and new units only used their own regional channels which did not match the channels of the Cologne fire service. Therefore incoming reinforcements had to be directed to waiting areas and from there were guided by local personnel to the assigned sectors, causing some delays.

These inhabitants, who would be affected by the flood first, were already warned of the flood at a water level of 7.30m KP on 21 December. Early that day, posters were hung in the streets with basic information on threats attuned to water levels, precautions to be taken, plus information for counselling, telephone numbers and radio channels. At the same day at 07.00 in the morning, Radio Cologne and the West German Broadcasting Service issued flood warnings and gave information that the water could rise above 10m and inundate parts of the city. Flood information was passed on to all media and frequently updated. The radio stations broadcasted hourly flood reports. The news papers gave detailed information on threats and advise on self-help for the population from 21 December onwards. Additionally loudspeaker cars of the police and fire service dispatched warnings in the morning, in the afternoon and in the evening of 22 December. As a consequence, inhabitants were warned of the inundation of the flood protected areas and the expected consequences 12 hours ahead of time.

During the floods, over 2,000 persons, members of the police, the fire service, the disaster response and relief organizations, and volunteers worked more than 100,000 hours. One of the major tasks was the construction of sandbag barriers to stem the flood and to make central facilities or private homes waterproof. To this end, 60,000 sandbags were used. In addition to the sandbag barriers, protective walls from steel or aluminium and 110 flood protection gates were used to protect the old town centre, tunnels, stores, pubs, restaurants and underground car parks. In many places the water was constantly pumped out of basements and underground structures to prevent major and lasting damage.

To maintain the lifelines in flooded areas and to support the pedestrian traffic in the flooded parts of the town, boats, rubber dinghies, high-wheeled trucks, and elevated walkways were utilized. The neighbouring houses in the old town centre in Cologne were linked by 3,400m of walkways. More than 40 boats were operating as "shuttle service". Like in Koblenz, the walkways were also used by disaster tourists.

In some areas, the obstruction of disaster response units by onlookers and disaster tourists on land and by small boats was disturbing. Due to the particularly narrow streets in the old town parts of Cologne, disaster tourists sometimes totally blocked traffic. As a consequence the response units could not always reach their destiny in time. In some cases barriers were erected against disaster tourists and even preventive measures against sabotage were taken. The areas along the mobile flood barrier which protected the old town of Cologne were cleared by about 60 policemen on 22 December. This action was taken after one policeman had seen disaster tourists who manipulated the fixing of the mobile flood barrier. In other places some disaster tourists destroyed sandbags with knives or they just pulled sandbags out of the barrier 'to give the flood a chance'.

In conclusion, the disaster management and the work of the response forces was efficient. Also the new sewer system for Cologne worked well. The communication problems with the disaster response services from outside Cologne were not severe and could be solved in the future by using common channels for events involving several regions. Only the protection level of 10m seemed to be too low to counter floods of this scale.

The 1995 response — Just thirteen months later on 25 January 1995, tremendous rainfalls, frozen grounds and thaw caused the next "once-in-a-century-flood". A flood wave went over the mobile flood protection barrier in Cologne. Almost 1,000 houses were flooded. While the flood situation aggravated in the neighbouring countries, France, Belgium and the Netherlands, the water levels remained stable in Germany on the Rhine river and its tributaries.

In Cologne, 33,000 inhabitants were directly affected. Many households experienced damage through rising ground-water levels and flooded cellars. Sandbags once again became a major, or in some places even critical device. They were used to rise the flood protection level from 10m in some places up to 11m, thereby preventing flooding.

The people living along the rivers in general and the residents of Cologne in particular had learned from the flood in 1993. They cleared the basements and groundfloors in advance and they set up sandbag barricades to draughtproof their doors and windows. Also they were better equipped than in 1993 to meet power failure and shortage of supplies, or they had arranged for accommodation with the family or friends for the days of the flood.

The officials and employees of the municipal departments and offices were also well prepared. Several hundred employees were available around-the-clock. Support was received from the different relief services and the German Army which assisted in the flood response for the first time in Cologne. Moreover several hundred police men-and-women maintained order and guarded flooded homes to prevent theft and looting.

The flood protection centre at the department for Sewage Treatment operated already during the minor floods beginning on 10 January 1995 when the water level rose above 4.50m KP, but only with limited personnel, since there was no immediate danger of a major flood. On 22 January at a water level of 3.70m, the Centre was again mobilized as rising water levels were expected. It started to operate at full strength on 24 January and coordinated the flood response measures of the fire service, the municipal services and offices, the relief organizations, non local disaster response forces and the Army. It was continuously informed about the execution of the assigned tasks and progress or problems. The Centre was occupied by ten representatives at all times while another ten employees took care of the telephone service, i.e. public information services and water level forecasts. At peak times up to 6,000 phone calls were received per day, mostly from residents of flood prone areas and journalists from all over the world. The open supply of information by authorities was appreciated by the media. The officials responsible for information services thought this form of media management to be quite positive and necessary.

The early warning period for the population and the authorities improved in 1995. The forecasts and information on the current and imminent development of the flood were highly reliable. Also the 24-hour period forecasts were of good quality. These forecasts proved to be decisive for Cologne and its residents in their efforts to avoid or reduce damage. Although the 1995 flood rose above the Christmas Flood of 1993 by 6 cm, the damage was limited to just half the costs of 1993. This was the result of precise forecasts and early precautions taken by the residents and the authorities.

At a water level of 4.50m KP the residents of flood prone areas were warned on 23 January at an early stage by posters, as before in 1993 (then at a level of 7.30 m). Inhabitants were informed about threats caused by the flood, measures of precaution, and means of access to the information services. This time an information leaflet was handed out to the households in the flood prone areas during December 1994. This leaflet gave detailed information on receiving current information and aid during the flood events, but also on how to improvise and to help oneself or others.

From the morning of 23 January onwards, Radio Cologne and national radio warned of the oncoming flood and extremely high water levels. This news was passed to other media and disseminated and updated every hour. First warnings were broadcasted via radio and TV to warn owners of vehicles not to park their vehicles in flood prone areas, or to remove them from the announced locations. This information was especially relevant to many campers who had their motor homes or caravans on endangered campgrounds near the bank of the Rhine river. In highly endangered areas, where the water passed over the flood barriers, loudspeaker-cars of the police and fire service supplied immediate warnings. Everything was done to supply at any time the necessary and reliable information to the residents and the disaster response forces.

During the 1995 flood no major problems occurred in the Cologne area except for the flooding of the old town centre. The authorities have decided to raise the basic flood protection level way above 10m. New calculations expect water levels of 11.30m KP for every 100 years and of 11.90m KP for every 200 years. Taking this into account, the city of Cologne started a Flood Protection Project which aims for different flood protection levels for different areas which can vary from 10.70m KP up to 11.30m KP.

In case of water levels above 11m KP in the present state of flood protection, Cologne would suffer a major disaster. Such a flood would necessitate the evacuation of about 100,000 residents out of about 965,000 residents for the enitre city. At this time Cologne is not prepared for such a mass evacuation. However, plans are being developed for such a contingency, since even with the improved and risen flood protection system, dikes may break. For this purpose the city of Cologne plans to set up a siren system along the Rhine river for an immediate areal warning.

6 Recovery and compensation

Tasks and responsibilities

The immediate recovery of the victims is the responsibility of the districts or municipalities, as long as they can handle it on their own and no declaration of disaster is issued for the whole state. With regard to rescue operations, evacuation, provisional accommodation and supply, clearance work and the reconstruction of public roads, the districts and municipalities are responsible as well.

The compensation for losses and reconstruction in the private sector exceeds the abilities and budgets of most of the districts and municipalities. No public or private insurance exists against flood damage, except for Baden-Württemberg. Up to 1 July 1994, the state of Baden-Württemberg had a compulsory and public insurance against elementary hazards, i.e. fires, storms, hails, floods (since 1960) and earthquakes (since 1971). The compulsory and monopolistic form of the insurance guaranteed compensation of risks, since it allowed a broad diversification of risks by type, while guaranteeing a sufficient pool of members. In 1994 the insurance was changed into a private and non-compulsory insurance owned for 98% by two former, in the mean time fused, monopolists who organized the public insurance before on state instructions.

Compensation is the responsibility of the State Ministry of Internal Affairs. The granting and approval of tax-deductible costs due to flood damage is the responsibility of the State Ministry of Finances. The granting of longterm credits for reconstruction in the private and business sector is the responsibility of the State Ministry of Economic Affairs. In addition, there are often special farmers recovery programmes run by the Ministry of Agriculture.

Damage inventory and compensation

An adequate and complete inventory of flood damages is difficult to make in general, but especially in Germany, since there is no single cost estimation method in the different municipalities and federal states. In several states only private damages and losses were registered, and only those above a certain value (above 3,000 DM for private and above 5,000 DM for business) covering only part of the total damage. In some places first estimates turned out to be exaggerated after investigation, while in other places the total damage did not become fully apparent until later stages.

Consequently, no exact numbers for the total damage done by the two flooding episodes can be given. Only enough estimates on the losses are available, except for North Rheinland-Westphalia, the only state which makes a complete inventory of all private and public damages. Both floods caused the same types of damages, economic losses and costs for the public and private sector (private households and business). The major types of damage, besides the tremendous ecological damages, were as follows:

Public Sector:

- damage to buildings, roads, bridges, waterways and their embankments and dikes;
- costs for disaster response units, fire brigades, police and other services;
- costs for evacuation, clearing and cleaning;

- financial aid for compensation of flood damages in the private and business sector.

Private Households:
- damage to buildings, furnishings, cars, etc.;
- costs for medical treatment as consequence of the floods;
- costs for repair and renovation.

Private Business (Trade, Transport and Farming):
- damage to buildings, fittings and equipment;
- loss of production, services and sales (business and shipping);
- costs for repair and cleaning;
- crop failure, loss of seeds, deterioration of the soil.

The losses for German states affected by the floods are estimated to total DM 1.3 billion. The states who suffered the major damage are: Rhineland-Palatinate (DM 650 million), North Rhine-Westphalia (more than DM 200 million, with DM 110 million just for the city of Cologne), Baden-Württemberg (more than DM 160 million just for insured buildings), Saarland (DM 100 million in total, 36 million for registered private losses) and Bavaria (DM 35 million).

Estimating flood damages in 1995 is even more difficult, since many authorities and officials have become very reluctant to give exact figures after the experience of the 1993 floods when many early estimates proved exaggerated and false. Additionally, figures concerning the financial aid to the private sector, granted by some states, are not available or not comparable to the figures of the 1993 floods. Many administrative provisions have since been changed, mostly towards more strict criteria for claiming in damage compensation. Weather and water level forecasts proved to be good. The inhabitants had learned from the year before and took the forecasts more seriously than in 1993. They cleared their homes early. This helped to prevent substantial losses. Just for Cologne the improved anticipation and preparation of the people helped to reduce the losses in January 1995 to DM 65 million, about half the losses of 1993.

Total losses in Germany for 1995 are estimated to amount to DM 500 million, compared to DM 1.3 billion in 1993. These figures do not include losses due to reduced tax revenues, necessary costs for long term reconstruction, improvement or new construction of dikes and other measures concerning flood protection in general. In Rhineland-Palatinate and Baden-Württemberg the flood of 1995 did not reach the water levels of 1993, and there was far less damage than in 1993. For example, Baden-Württemberg flood losses in 1995 are estimated at about DM 50 million (plus DM 20 million due to storms) compared to DM 160 million in 1993 just for insured buildings. In contrast, North Rhine-Westphalia was hit more severely by the flood of January 1995 than in 1993, but due to the better preparation, losses can also be expected to be significantly less than in 1993. Cologne presents a similar example (1993: DM 110 million; 1995: DM 65 million).

The losses due to flood damage of the waterways and structures relating to the waterways are about DM 12 million for 1993, DM 7.5 million of which concerns the Saar and Moselle river. The Rhine and the other tributaries suffered relatively minor damages. In addition during the 1993 floods the Moselle river was not navigable for 12 days, the Neckar and Saar for 9 days and the stretch of the Rhine from below the confluence with the Moselle up to the Dutch border for 7 days. Altogether these opportunity costs totalled about DM 50 million.

Ecological damage occurred in form of sedimentation of flooded grounds, both in agricultural and nature reservoirs. Since the Rhine and many of its tributaries are joined by industrial plants, such as chemical plants with poisonous sewers, the river may carry harmful chemicals which can cause severe damage by poisoning the ground and upsetting biotopes. Some of the spoiled agricultural grounds cannot be used any longer after flooding due to the high contents of harmful chemicals in the plants. Also sensitive biotopes may be set off balance by the chemical poisoning as well.

Sedimentation is also an expensive and time consuming effect of the floods. Especially the areas, flooded with water without or just little current were covered by several centimetres of sediments after the flood retreated. Since the sediments dry fast and form a solid layer, they have to be removed immediately when the water is retreating to avoid extremely expensive and labour intensive removal at a later point of time. The cleaning of the public streets and places in Cologne costed about DM 1 million. Furthermore the spilled heating oil caused considerable costs. In Cologne almost 2,000 tons of this water-oil mixture had to be removed and disposed of by specialized companies.

Flood insurance and compensation in Germany: a closer look

Except for the state of Baden-Württemberg no flood insurance in Germany for business and residential properties exists. For the Christmas Flood of 1993 the total losses are estimated to be more than DM 1 billion, with an average loss of DM 16,000 and with the largest single insured loss totalling DM 7 million. The burden of these losses falls primarily on the owners of private business and property. Compensation paid to individuals for damage and losses generally makes up for only a minor part of total losses. People mostly relied on themselves and the help from family, friends or neighbours. They accepted damage without much complaint.

The different states generally granted depreciation provisions to the victims to write off the costs of the flood damage. But there are no tax-deductible expenditures for private flood protection measures. In addition, credits for reconstruction were granted at low interest rates with repayments spread over several years. The compensations for damages were only granted if the damage surpassed e.g. DM 3,000 for private property or DM 5,000 for business. Minor damages had to be paid for by the people themselves.

The example of Koblenz shows that most of the damage and losses occurred in the private sector. Estimations of the total damage in Koblenz range from DM 150 to 200 million. The losses just for private households are estimated at about DM 57 million. At the same time, only about DM 6.7 million worth of damage compensation claims have been put in by the people who suffered losses. The total damage for the private sector in Koblenz was split up into 16 classes. An evaluation clearly showed that people with minor damages were either not entitled to compensation claims, or when they were, most of the time abstained from claims. The smaller the losses, the less likely people claimed compensation. In all, less than 10% of the people who suffered damage and losses applied for compensation, while more than 90% did not.

Data regarding insurance and reinsurance clearly show that there is a worldwide trend towards increasing economic damage caused by natural disasters. Long term and costly measures are to be taken against riverine floods, but these measures should pay off in the long run compared to the longterm expenditures on potential flood damage. In Germany, the damage caused by riverine floods is mostly paid for by the victims. The government pays only little compensation and grants only general tax deductions. Since there is no insurance against the risk of flood damage, substantial personal and economic risks fall to the victims of floods. If "once-in-a-century-floods" occur every few years in the near future, and if government and authorities do not change their compensation policies, floods may ruin the lifes of many people living and working along the rivers. One has to expect an increasing gap between the political and public perception of the causes of riverine floods and the management of flood damage by the authorities, as well as an increasing annoyance of the public due to the passiveness, negligence or even ignorance of the effects by the authorities.

Notes

1 The German case was included at a relatively late stage of the project (August 1995). Therefore no data concerning the people affected and the management of the riverine floods in December 1993 and January 1995 could be gained by first-hand observation. We conducted a series of telephone interviews with administrators at the different levels of administration and decision makers at the communal level and with representatives of the different disaster

relief services, reinsurances and hydrological and meteorological research institutes. Furthermore, public evaluation and investigation reports, which were made on behalf of the communities affected, the states or the federal government (solely hydrological and meteorological reports) were analyzed. These reports all provide highly aggregated data. We held additional interviews with decision makers in Cologne and Koblenz to obtain sufficiently detailed information on the actual process of decision making and crisis management. These interviews provided additional insight into the local problems of decision making and disaster management. Finally, reports of the major reinsurances were evaluated and a content analysis of the media coverage of the flood events was made.

2 State water laws maintain the principle of private water property, except for the state of Baden-Württemberg.

3 DWD utilizes data from a nationwide network of weather stations and METEOSAT-satellite pictures for the development and improvement of weather forecasts and forecast-models.

4 According to communal law, the Lower Disaster Response Authority is either attached to the district *(Landkreis)*, to the administrative area *(Regierungsbezirk)* or to the municipality *(kreisfreie Stadt)*. For example, in the states of Rhineland-Palatinate and North Rhine-Westphalia administrative areas can consist of several districts. The municipalities *(kreisfreie Städte)* like Cologne and Koblenz are independent corporate bodies on the level of districts or administrative areas. The community *(Gemeinde)* is the basic administrative unit at the local level. It provides the fire brigade(s) and the local police.

5 Since the actual organization of the Disaster response in the different states is basically similar, but differs in detail and in name, the specific details for Koblenz and Cologne are given in section 5.2 and 5.3.

6 Not included are the water level stations of the WSV.

7 Additionally the Staff HVB experienced problems in gaining up to date information on the meteorological and hydrological situation, because they were informed via the police by means of low quality photo-copies or faxes converted to BTX-pages which were unreadable. This problem was solved after the 1993 flood by using a professionally organised fax-distributor, which in 1995 was able to supply at short notice the latest data and forecasts of the meteorological situation, and the relevant water levels of the Rhine and Moselle and its tributaries.

8 Based on interviews with officials affected.

Chapter 6

FLOOD RESPONSE DYNAMICS: LOCAL RESILIENCE AND ADMINISTRATIVE FLEXIBILITY

Uriel Rosenthal, Michel Bezuyen, Menno van Duin and Paul 't Hart

1 Introduction: a constrained comparison

This concluding chapter contains a comparative review of disaster management during the floods of 1993 and 1995 in France, Belgium, the Netherlands and Germany. The emphasis will be on the country experiences with respect to risk communication, the organization of disaster response (including evacuation decision making), and damage compensation. The chapter will conclude by reflecting upon the 1993 and 1995 flooding episodes as occasions for policy-oriented learning in flood and disaster management.

We emphasize the limited scope of the conclusions drawn here. In many respects, the cases studied here constitute a rather specific sample, and it would be inappropriate to draw generalized conclusions about 'disasters' and 'disaster management' from them. To begin with, there are substantive differences between floods and other kinds of natural disasters, and a fortiori between floods and man-made disasters. Also, the sequence of two highly similar flooding episodes within eighteen months made for a specific constellation of facts, expectations, planning and learning processes. Simply put: in all of the countries studied the disaster management process in 1995 was shaped in large part by the recent experience of major flooding problems in the 1993 case. To treat them as a separate case would be a major methodological error. Moreover, despite the many similarities between the four countries, one has to take into account a number of equally striking differences. Both the situational context of the floods and the institutional context of disaster management differ in important ways. Finally, although the frame of reference in this analysis derives from the body of knowledge of pure and applied disaster research, the differences in severity, size and impact of the floods in the cases examined here may raise some questions about the appropriateness of 'disaster' as the dominant conceptualization for at least some of the events and developments in the four countries. In terms of the social disruption caused by the rising water, the 1993 and 1995 floods were sometimes more a disturbance than a major calamity for some communities and regions. Consequently, analysts studying responses to the floods should expect to see marked differences in the kind and intensity of collective behavior and administrative crisis management commonly associated with full scale disasters.

2 Risk communication: flood forecasting and warning

Flash floods and riverine floods

In the domain of risk communication in particular it is important to distinguish between the challenges entailed by flash floods and riverine floods. Flash floods involve a sudden increase in water quantity and thus belong to the specific category of sudden onset disasters. Their very speed and magnitude almost ensure that flash floods assume critical proportions instantaneously. Our sample of the 1993 and 1995 floods includes one such flash flood, e.g., in the French region of the Vaucluse in September-October 1993. Riverine floods in contrast are characterized by a slow rise in river levels. The threat to the population develops gradually, although communities may still be caught by surprise due to the misreading of forewarnings or sheer negligence. Riverine floods are the dominant category among the 1993 and 1995 floods in the four countries, with multiple cases for each country described in this study.

The distinction between flash floods and riverine floods determines for a large part the context and the specifics of forecasting and warning. The sudden onset of flash floods makes it virtually impossible to convey comprehensive and timely warnings to both the local authorities and the population. During the flash flood in the Vaucluse communes of Valréas and Bollène, there were no official BRAM or ALARM pre-warning messages. Nevertheless, local officials and volunteers, more in particular experienced firemen, made an adequate assessment of the bad weather in other areas, issued proper warnings on the spot and helped their communities to react in time, testifying to the crucial importance of unofficial flood warning mechanisms at the grass roots level.[1]

The slow onset of the riverine floods in Belgium, Germany and the Netherlands provided the authorities and the population with considerable lead time. For example, it took about four to five days for the flooding water to arrive in the Netherlands after the initial floodings has occurred in upstream countries. However, the time available was not always used effectively. Particularly in 1993, there were many examples of sluggishness in the dissemination of risk information, both from one public agency to the other and from authorities to the public. Also, there was a much greater reluctance among the inhabitants of the areas at risk to take seriously the warnings that were issued. Both the authorities and the public in France appeared to be more used to floods. There may not have been the kind of fully developed disaster subculture and/or integrated forecasting and warning system that may be found in areas where disasters are endemic and frequent, but it is clear that especially in 1993 there was a much quicker response to signs of impending flood problems in French communities than elsewhere.[2] In part, the quick response to the Vaucluse flash flood may have been due to the vivid memories of the calamitous flash flood in nearby Vaison-la-Romaine in September 1992. However, also in 1995, mayors and the local agencies in the northern part of France were alert enough not to wait for the official warnings from the specialized state agencies to start issuing public warning and take effective measures.

The role of the meteorological and hydrological services also varies with the type of flood.[3] To the extent that, confronted with flash floods, they could benefit from early warning input from state agencies, French local authorities appeared to rely essentially on weather forecasting. Faced with riverine floods only, Belgium, the Netherlands and to a lesser degree Germany tended to rely primarily on their hydrological services. Consequently, they were more dependent on hydrological information from neighboring countries. Within these countries, highly detailed information about current and future water levels proved to be a key factor in disaster preparation and response.

Deciding upon the timing of pre-warnings proved to be rather difficult. Wary of the cry wolf syndrome (see chapter 1), agencies were hesitant to give warnings too easily. They wanted to gain as much certainty as possible about the impending danger, even when the kind of information needed was not always available. At times, this rigid maintenance of professional norms and standards had a paralyzing effect.

Interorganizational communication problems

All four countries experienced problems with forecasts and communication deficiencies in relaying forecasts to authorities and rescue services. Four types of problems were particularly prevalent. Firstly, many emergency services and authorities complained about the content and the timing of flood announcements. In France, for example, the Vaucluse region did not receive any warning signals from Météo-France, thus resorting to other sources of information on the coming flash flood. Likewise in 1993 in the Netherlands, the region around Maastricht was plunged into a flood that should have been foreseen and warned for virtually without prior notice.

Secondly, there were remarkable discrepancies between forecasts made and the actual situation experienced by agencies working on the ground. In the Netherlands, the department of Public Works' forecasts appeared to be inconsistent with predicted heights and peak-level times. During the floods in the French Ardennes, the forecasts of the SAC flood announcement service were criticized by a number of mayors who regularly noticed major discrepancies between SAC forecasts based on meteorological data, and the weather and water levels they were confronted with locally.

A third problem was essentially stylistic, yet a pivotal constraint of the effectiveness of the warning process in all of the countries: the technical and untransparant vocabulary used by meteorological and hydrological services. Flood bulletins of these services were difficult to read and interpret for laymen, causing misunderstandings among local authorities and emergency services. These problems eventually produced serious tensions between the agencies responsible for issuing warnings and their clients throughout the governmental system.

Finally, the transnational dimension of the floods brought about conspicuous delays and other communication problems. Especially in 1993, communications between France, Belgium (even between the Flemish and Walloon regions) and the Netherlands were not very effective. Initially, the French authorities and specialized services did not convey their information about the water levels on the Meuse, Schelde, IJzer and their tributaries to Belgium and the Netherlands. As it turns out, they were supposed to exchange information with their counterparts in the neighboring countries only when the situation in those countries would have become really urgent. Likewise, the communication and dissemination of information between Belgian and Dutch authorities were rather onerous. A complicating factor here was the calibration of measurement standards. For instance, 47.32 TAW in Belgium equalled 44.65 meters above sea level in the Netherlands. The 1993 transnational communication problems were not the result of bad will or international politicking. They mainly derived from the absence of common orientation and a general lack of organizational routines encouraging water and meteorological authorities to take a transboundary view of flooding problems. The 1993 experiences helped change this, at least temporarily, and in 1995 communications and information sharing on weather conditions were considerably more effective.

Warning the public

A vital aspect of the warning process was the communication of risks, impending threat, and finally manifest danger to the relevant populations. The four countries used a wide variety of public warning modes: loudspeakers, press releases, house-to-house letters, national television and radio networks and, of rapidly growing importance, local and regional broadcast stations. The selection and use of specific methods depended strongly on the time available for proper and effective warning. In Valréas, it was simply too late to send a vehicle with loudspeakers through the streets. Authorities and emergency services in Germany,

Belgium and the Netherlands had more time and could therefore use more advanced and diversified methods, including the involvement of the mass media. Nevertheless, as noted above, the longer lead time in Belgium and the Netherlands did not always ensure that the warning process was effective. In 1993, a considerable part of the population in the stricken areas of Belgium and the southern part of the Netherlands got the message only when the water had already risen to its uppermost levels. If warned at all, they received the relevant information when it was too late to protect their houses, animals, businesses, and personal belongings – their life and limbs were never genuinely at risk.

The floods of 1993 and, to a lesser extent, 1995 once again showed that the soft spot of public warnings is how to convince the public that the impending danger is real if the pre-existing social context is one in which flood risks are not deeply embedded in the public consciousness. In all four countries citizens and firms tended to pay insufficient attention to the announcements made by the authorities. In some areas the sheer frequency of flooding had prompted people to rely upon their own intimate knowledge of risks and dangers. Both in the Vaucluse and in the Ardennes a sizeable proportion of the population engaged in forms of 'DIY risk assessment' which easily gained mouth-to-mouth credibility.[4] Official forecasts and forewarnings could not match the strength of this kind of inter-personal communication, confirmation, and, more particularly, negation and denial. In addition, the scope and impact of the 1995 floods caught many communities and citizens, just like the authorities and the emergency services, by surprise. Flooding had occurred in the past, but only on a small scale. When water levels rose in 1993 in Germany, Belgium and the Netherlands, a major flood was generally viewed as a once in a century phenomenon. Nobody expected a new wave of floods within only two years. Furthermore, it proved difficult to convince the population in these countries of the magnitude and severity of the current water problems, since people were more inclined to judge the warnings in view of earlier flooding experiences which had been so much less severe.[5]

3 The organization and implementation of disaster response

Institutional settings and strategic policy decisions

The official responses to the floods were shaped significantly by the existing institutional structures of disaster management, which differed markedly between the four countries. In France the state's role in disaster management is focused on the level of the *Prefecture*. According to the 22 July 1987 law, the Prefect coordinates the public rescue services. The so-called Defence Zones provide an administrative and organizational layer between the State and the departments when it comes to matters of civil safety. The Zone Prefect has the right to ascertain, allocate, centralize and prioritize the request from other prefects in the zone. On the local level, municipal representatives are responsible for setting up crisis and emergency units such as coordination centers and operational command posts.

In Belgium, the main thrust of the administrative organization of disaster management is the multi-phase model. The model was introduced in 1974 and has been tailored to the governmental authorities responsible for disaster planning. The model distinguishes four phases. During phases 1 and 2, the mayor is responsible for emergency action and is supported by the municipal crisis center. In the next phase (phase 3), the provincial governor assumes overall responsibility for the rescue operations with municipal crisis centers concentrating on assisting the provincial coordination center. In phase 4, the federal Minister of the Interior assumes prime responsibility and sets up a governmental crisis and coordination center. The model is seen as the guiding principle in up-grading the scope and leverage of political and administrative intervention in accordance with the scope and severity of the disaster.

In the Netherlands primary responsibility for disaster response falls on the mayor. This decentralized approach is essential to the way disasters are handled in the Netherlands. Whenever a disaster extends the boundaries of the municipality, a regional coordination center is set up. This center performs the function of liaison between the municipalities and the province. Provincial or national authorities can intervene and issue guidelines or instructions whenever the mayors face coordination problems. Disaster legislation also allows for provincial or central intervention in the case of urgent reasons of common interest.

In Germany responsibilities for disaster response concentrate on the level of the states *(Länder)*. Federal legislation obliges the states to make preparatory arrangements for emergency management. The states have delegated their actual disaster management tasks to lower authorities. Depending on the communal constitution, this can be done by a district, administrative area or municipality *(Kreisfreie Stadt)*. In practice, therefore, the German system is highly decentralized, but with a more self-evident structure for upscaling than in the Dutch case, where local autonomy in this domain has long historical roots and has been zealously protected by local government, in particular the mayors.

Strategic decision making: formal structures and informal practices

The most striking difference between the different countries with regard to their actual response to the 1993 and 1995 floods is the political and administrative level on which operations were coordinated. To achieve a proper intergovernmental division of labor and balance of power is no easy task.[6] In Belgium and France, respectively, the provincial and the departmental level were dominant. The provincial governor in Belgium and the Prefect in France played an important role in the response to the floods. In Belgium, the coordination of relations between the provincial level and the municipalities required a lot of attention in 1993 as well as in 1995. If only for that reason, the provincial governors considered it absolutely necessary to keep federal intervention within limits. Involvement from the federal level would have imposed an additional burden upon the coordination of the rescue operations which by itself had already become quite complex. Federal coordination would anyhow have been a difficult affair for the simple reason that there was no overall plan to handle these situations.

In France, the active role of the *Prefecture* in flood management in 1993 and 1995 matched general administrative practices. Nevertheless, it is hard to underestimate the actual importance and leverage of the local authorities in many parts of the stricken areas. Local presence and the quality of information gathering and dissemination within local communities and towns forwarded them with an intimate knowledge of what was going on, and for that reason gave them a strong position in determining the course of action.

In Germany, the bulk of the response was made at the local and sub-local level. In the Netherlands, the distribution and allocation of responsibilities between the national, provincial and local levels of government were more complex. Although the local authorities were responsible for the operations, provincial and, especially in 1995, national authorities took an active stance. In 1993, the southern part of the Netherlands was struck by the flood. This led to coordination efforts on the regional and provincial level. In 1995, three provincial governments were actively involved in addition to Limburg. The bad state of some of the dikes in the central parts of the country caused much media attention and public concern, creating a policy issue that went above and beyond the local level. Furthermore the geological situation of a greater part of the area created anxiety about the so-called 'bathtub effect' of water flooding into the extensive flat Hinterland surrounded by river dikes. Intergovernmental coordination up to the national level appeared to be necessary to meet the scale of the evacuation operation that ensued. Eventually, the Minister of the Interior became directly involved in decisions concerning the timing of the return of the people to their homes. Intriguingly, all this took place despite the fact that formal responsibility regarding public order and safety remained at the local level, and indeed quite a few mayors used their prerogatives to reverse evacuation decisions made by higher level authorities.

Strategic decision making: local dominance?

This last observation highlights a more general point. The flood experiences showed once again the limits of the 'centralization thesis' which holds that crisis decision making in administrative systems inevitable entails a concentration of power at higher levels.[7] Various alternative response structures emerged during the floods. As we shall discuss below, many strategic decisions were made at the local level. In part, this was because of preexisting structures for decentralized disaster response. At the same time, however, during the French flash floods in particular but also following the surprise riverine floods in 1993 in Holland, there were also situations of sheer situational dominance, where extreme time pressures forced local authorities or even operational personnel to take on the spot decisions. At later stages during the floods, there were indeed attempts to centralize decisional authority in some countries, but those were sometimes actively resisted by local or regional administrations, with decisions emerging from intergovernmental bureaupolitics rather than uniform leadership by one administrative layer or the other.

Despite the administrative and organizational differences between the four countries, local leverage was a recurrent phenomenon during the floods. This was clearly manifest during the earlier stages of the floods. It speaks for itself that the first entities to be hit by the floods were local communities and that the rescue operations were begun and supervised initially at the local level. Municipal agencies had the significant advantage of 'local presence', e.g., a close familiarity with intricate local situations and problems. Partly because of this, they also enjoyed both general and situation-specific support among the local people.

In France, both in Valreas and Bollène rescue operations were organized primarily at the municipal level. In the Ardennes, local authorities and municipal agencies also played a pivotal role in rescue management. In Belgium, especially in the province of Namur, the mayors were dominant in the decision making concerning the evacuations. On several occasions the provincial coordination committee gave an order to evacuate, only to find the execution of the order was delayed or blocked by the mayor involved. In the Netherlands local, provincial and national authorities took part in a wide variety of coordination arrangements. Even in some quite small communities a mayor who really insisted on exerting his formal powers, could successfully challenge the disposition of provincial and even national authorities to intervene. In Germany, disaster response was more and more decentralized in 1995. The 1993 flood had shown that it proved to be more efficient to decentralize the command structure of the Disaster Response Force and to transfer responsibility and decisional freedom to on-site officials. As a result the federal, state or communal level proved to be of minor importance in shaping the 1995 flood responses.

Operational decision making: organizational pluralism

One of the most characteristic features of disaster management invariably are the substantial numbers of organizations and agencies involved.[8] In each of the four countries, many organizations played a more or less important role: local, regional, provincial and federal agencies; public and private organizations; professional and volunteer organizations. Although the various countries displayed interesting differences in the specific nature of organizational leverage and inter-organizational arrangements, there were also some intriguing similarities. We briefly review the experiences with various categories of operational agencies.

Emergency services: fire brigades at the helm — The most remarkable observation is the important role of the fire services in all four countries. They coordinated the activities of most operational services – the fire service chief (commander) being responsible for the coordination of the activities on the part of the different emergency services. In general, the fire services dealt adequately with their task, even though inter-agency coordination often involved highly sensitive mutual recognition of fairly recently established competencies and responsibilities. If there were problems to be addressed, they tended to bear upon the suddenness and severity of disaster circumstances which put high pressure on the fire services and the other emergency services alike.

Armed forces: a backup support system — The floods also challenged the civilian function of the armed forces. During the evacuations caused by the flash floods in 1993 and the riverine floods in 1993 and 1995, the military became engaged in various parts of the rescue operations. German, French, Belgian and Dutch soldiers worked alongside personnel from the respective civil emergency services. Their tasks included the filling of sandbags, the escort of the evacuation and the deployment of equipment. The main problem turned out to be the civil-military coordination, especially in Belgium and the Netherlands. Prior contacts and networks across the two domains had been limited. In 1995 civil-military coordination was noticeably smoother. Except for France, public authorities showed considerable restraint in their deployment of the armed forces. It had been a long time since military personnel had last been asked to take an active part in disaster relief. Many authorities felt somewhat ill ease with the thought of large-scale military operations in a purely civilian emergency. Furthermore, they were not familiar with the impressive assets, in terms of personnel, equipment and logistical capacity, of the armed forces. When, after ample consideration, local authorities called upon the armed forces for assistance, it took quite some time before such assistance materialized. The internal military procedures and the various layers of civil-military consultation and coordination were time-consuming.

Civil defense agencies: on the wane? — In the Netherlands, the civil defense organization was abolished in the eighties, giving way to a more prominent role for the fire services. In the other countries, however, a civil defense organization still exists. On various occasions during the floods there were considerable tensions between the civil defense and the emergency services. In particular in the province of Namur (Belgium), the position of civil protection services was strongly criticized. Some fire service units refused to cooperate further with relief workers belonging to the civil protection services. While there may have been bureau-political considerations at work here – in Belgium in particular interorganizational strife between different tribes of 'men in uniform' is deep rooted[9] – it may also reflect the diminishing practical significance of the civil defense organization and the growing importance of the fire services.

Voluntary organizations: a massive turnout — The floods also solicited a massive influx from volunteers. Overall, the involvement of voluntary organizations and volunteers was impressive. In the local context, citizen groups, voluntary associations and intermediary organizations played an active part in the disaster response. Voluntary assistance could relate to a variety of positions and situations, however. For instance, there was a striking difference between the lack of experience on the part of millions of volunteers in Germany and the relatively high level of training and experience of volunteer firemen in the Netherlands, the latter being considered as an absolutely necessary part of the Dutch fire services system. Having said this, the massive assistance of volunteers sometimes created distinctive interorganizational problems. Voluntary activities often came as a sheer surprise for authorities and emergency services harboring a strongly state-centered policy perspective. There were sometimes predictable tensions between paid professionals and 'amateur' volunteers, the former insisting on playing a leading role and trying to relegate the latter to minor support duties. In other cases however, cooperation was productive, up to the point of formalization of the role of voluntary organizations in the emergency response. For example, around Koblenz (Germany), an emergent communal self-help group was eventually integrated in the command structure of the disaster response.

Transnational cooperation: an emerging asset — During the 1993 as well as the 1995 floods, the countries involved took several initiatives for mutual operational assistance. Firemen and other emergency personnel rendered important services to their colleagues abroad. This kind of assistance was particularly effective in those very instances where the emphasis was on informal, operational, situation-specific cooperation on the spot. Contrary to the gist of some of the elevated political and administrative thinking on transnational cooperation, well-defined and pointed arrangements among adjacent border areas, especially Euregional cooperation schemes, worked much more smoothly and effectively than large-scale over-all agreements between national governments of the countries involved.

Authorities, rescue services and the public

The organization of disaster response can never be a fully administrative affair. Whatever action the authorities and agencies choose to undertake following a disaster, they have to take into account spontaneous social processes at the community level. This partly constrains the scope of government action, but it partly also defines the target of policy interventions. Here we review two well-known collective behavior phenomena: self-regulating public behavior and disaster tourism.[10]

The self-regulating society — Disaster researchers are well acquainted with a number of so-called disaster myths. These popular beliefs about human responses in disaster are not based on scientific knowledge but on conventional wisdom, which often becomes reinforced by the media. Examples are the expectation that people will panic in a disaster; that there will be widespread looting; that many people are very eager to evacuate in a disaster situation; and that they will scramble for access to public shelter facilities. In general, there is a popular belief that people threatened or hit by disasters will act irrationally. These myths were also present during the response to the floods, guiding the expectations and actions of authorities and operational services. Time and again, they were surprised that many people kept very calm and were not eager at all to evacuate. Likewise, crime proved to be the least of their worries, and in most cases public shelter facilities remained almost empty.

Taken together, these examples illustrate one of the most striking phenomena during the floods: the generally rational and effective stance towards the disaster taken by the population in the stricken areas. When and where early warning time was absent or inadequate, local communities tended to reveal a self-regulatory capacity that was functional instead of falling prey to the kind of mob panic that many authorities continue to fear. In fact, this kind of self-regulation indeed was one of the main factors explaining the success of some of the larger evacuation operations undertaken in Holland and Germany in particular. The relatively limited role of the rescue services in the escort of the evacuation, together with the significant numbers of people who stayed with relatives and friends during the evacuation period, are further witness to this observation.

However manifest and impressive this self-regulatory attitude of the population may have been, it also had some negative side-effects. Some people refused to evacuate because they felt adequately protected in the upper floors of their houses. In some instances, people waited so long that evacuations really had to take place under very difficult, if not life-threatening circumstances. Such instances of non-cooperation not only cost substantial extra time, but they also involved extra risks for both the inhabitants and many rescue workers.

Disaster tourism — All country studies provide marked examples of so-called disaster tourism. These include the hampering of disaster operations by onlookers in Namur and Cologne, both in 1993; individuals ignoring signs to enter forbidden areas in the Ardennes, 1995; media-instigated and commercially exploited forms of organized disaster tourism in Koblenz, 1993; and finally active sabotage of flood barriers in Cologne, 1993.

The arrival and sheer numbers of disaster tourists hindered rescue workers and required policymakers to take countermeasures, such as sealing off endangered dikes and strategic routes. The implementation of these measures required the deployment of additional police on public order assignments in and around the flooded areas.

In the wake of the floods, much was made of the disruptive impact of disaster tourism, and media were presenting the phenomenon as a growing trend that was becoming an increasing nuisance to disaster managers. This representation needs to be put into perspective, however. To begin with, disaster tourism is as old as disasters themselves. Authorities should not be surprised by this kind of behavior, but should rather be prepared for this phenomenon. Increased communication and mobility have only amplified the problem, not created it. Also, it is important to distinguish between different components of the mass convergence on the disaster site. Each of these flows has its own rationale and modus operandi, and each poses a different set of potential disaster management problems.

One flow consists of agencies and volunteers mobilizing and calling for on-site duty. Parts of this flow are intended and planned for, whereas others are spontaneous and unexpected. For example in the Netherlands, specific calls for assistance of fire personnel throughout the country brought some surprises of their own. Some fire brigades offered their help spontaneously, not waiting for requests. A second flow involves friends and relatives of possible victims seeking information about and contact with their loves ones. Thirdly, the seemingly ever increasing army of journalists and news teams that assault the site of major disaster constitute the fastest growing contributor to mass convergence. In the literature on disasters and the mass media, the sheer number and highly strung activism of crisis reporters is acknowledged as a major logistical and PR challenge to policymakers and operational services alike.[11]

Finally, then, there are the curiosity seekers and 'tourists'. These are of two kinds: ordinary citizens and VIP's. Conventional treatments of disaster tourism concentrate entirely on the former, but recent studies indicate that it would be wise to pay more attention to the latter as well.[12] In several of the cases in this volumes, political and bureaucratic elites (Royalty, ministers, high-level civil servants) visited the stricken areas in large droves. Of course, such visits have important positive functions. They may provide moral support to a community that is down and out. They may also underscore official promises of financial and material support. At the same time, however, on-site visits by elites may require a lot of attention and energy from authorities and operational services already severely burdened by the ongoing disaster operations. In the worst case, search and rescue operations have to be interrupted to accommodate the security needs or media paraphernalia that are part and parcel of these elite visits to disaster sites. Moreover, sometimes such visits are primarily self-serving acts of political dramaturgy and governmental propaganda.[13] If so perceived locally, they may actually put off rather than serve the needs of victims and field workers.

4 Managing evacuations

Evacuation is one of the most important, if not critical, aspects of disaster and flood management. It is sometimes claimed that evacuations should indeed be considered as a defining characteristic of full-fledged disasters. This may be an exaggeration, but it is true that the prospect of evacuation forces authorities and emergency services and citizens to make unmistakable and highly consequential decisions.

Wherever they take place, floods solicit concerns and indeed decisions to evacuate people from threatened locations and areas. Some evacuations will be government-instigated and steered, others will be the result of more or less spontaneous collective behavior within the community. Evacuations may follow from a protracted process of calibration, or they may occur spontaneously. With the benefit of hindsight, one may place evacuations on a continuum that runs from bitter necessity in a major emergency to the irrational overreaction to a not so threatening extraordinary situation.

The floods of 1993 and 1995 in the four countries under examination are no exception to the rule that flood management is intimately linked up with the predicament of evacuation. In all four countries evacuations took place, varying from a carefully planned mass evacuation of about a quarter of a million people in the Netherlands in 1995 to small-scale, spontaneous and utterly necessary evacuations during the 1993 flash flood in France. The evacuations were relatively short-lived. Some evacuations did not take more than a couple of hours, others lasted longer. In some cases people were obliged to leave; in others, citizens left their homes on a voluntary basis. Most of the evacuations studied here were unpleasant and messy for those affected, but generally fell short of the kind of tormenting experience described in international disaster studies covering other floods and life-threatening emergencies.[14]

The findings on the evacuations of 1993 and 1995 are presented applying an evacuation typology that uses two main variables: the degree of threat and the utilization of available warning time.[15] The threat levels entailed by the 1993 and 1995 floods varied from relatively high to relatively low. With regard to utilization of available warning time, three types of situations were manifest. Firstly, there were cases of a total or nearly total lack of warning time and thus a complete surprise for potential evacuees. Secondly, in some cases there was sufficient warning time but no use of it was made to ensure a prompt evacuation. Sometimes authorities simply lacked the technical means to effectively inform and assist the people to be evacuated; yet in some cases they also deliberately chose to warn people at the last possible moment in order to prevent panic. Thirdly, there were instances where full use was made of the available warning time, enabling the authorities and the affected population to prepare step by step for the coming evacuation. In table 6.1 the cases featuring evacuations examined in this study have been placed into the cells of the typology.[16] We shall discuss each cell in turn.

Ad 1 Almost by definition, flash floods such as those of Vaison La Romaine in 1992 and the Vaucluse in 1993 strike suddenly, unexpectedly and with great speed. Stormy weather conditions aggravate the assault of huge quantities of water and evacuations generally take place under extremely bad circumstances. By the sheer force of sudden threat, most of these evacuations take place spontaneously. This does not imply that they are by definition acts of panic and irrational behavior. In contrast, they are better understood as perfectly sensible responses by individuals to extreme circumstances. The disorganized character of these evacuations derives from the fact that large numbers of people have the same thought and start moving at roughly the same time, putting pressure on the available infrastructures and reinforcing the uncertainty and urgency felt by the evacuees. In the cases studied here, quick action by the authorities effectively managed to set in motion rapid evacuations in life-threatening situations. For example, in Valreas local officials, mayors and firemen in particular, showed a keen alertness to the impending danger and were able to organize precautionary evacuations of campsites which, using the 1992 flash floods as an historical analogy, they had identified as the most vulnerable targets.

Table 6.1 Types of evacuations in 1993 and 1995

Availability and utilization of warning time / Severity of threat	Absent or minimal	Sufficient, but not used	Sufficient, and used
Life-threatening disaster	1 France (Vaucluse) 1993	2 Netherlands (Gelderland) 1993	3 Netherlands (Gelderland) 1995
Not life-threatening disaster	4 no cases	5 France (Givet) 1993 Belgium (Limburg) 1993 Belgium (Namur) 1995 Netherlands (Limburg) 1993	6 France 1995 Belgium (Limburg) 1995 Netherlands (Limburg) 1995

Ad 2 This category focuses on two flood constellations: flash floods where, after all, something could have been done to arrange for some sort of orderly evacuation; and riverine floods where for a variety of reasons relatively ample warning time was not used to the fullest. In the Valreas case, the deficient communications from Météo France would have triggered the former scenario, had it not been for the effective monitoring and improvisation at the local level. The second scenario pertains to the case of the Dutch region of Gelderland in 1993. In 1993 the authorities apparently did not utilize their resources, including facilities for effective warning, to get a substantial part of the population prepared for evacuation that could well have become necessary a few days later

(but later transpired not to). This comes out most clearly if one compares it to the process in 1995, where roughly similar indications triggered government-led mass evacuations.

Ad 3 Floods may involve a steady rise of the water level up to a certain height, thus inundating houses, buildings and infrastructural facilities; the floodings of the river Meuse and parts of the Rhine in Germany are a typical example. Of course, on an individual level this may cause serious inconveniences and even risks but on the societal level it will most likely be a relatively well manageable situation, with the main problems caused by the indirect effects of flooding such as the disruption of energy facilities or transportation problems. An entirely different situation, however, pertains to the impact of rising water levels on water works such as the river dikes. Protracted periods of high water levels may bring about their gradual or sudden collapse and produce a disaster of catastrophic proportions in the Hinterland. In 1993 and in 1995 the threat to the central part of the Netherlands emanated not so much from the rising water level of the Rhine as such, but from the insecure and unstable condition of some of the dikes in this area. The danger of a collapse or breach of weaker parts of the river dikes induced the authorities to evacuate large numbers of people. They were led by the knowledge that in some villages the breaching of a dike would make for a rise of the water level by more than 5 meters within 24 hours.

The situation in the province of Gelderland in 1995 was potentially life-threatening, but the authorities had enough time to make plans and to alarm and warn the population involved. This helped to shape the conditions under which the evacuations could proceed rather smoothly. There was enough time to evacuate, and both the authorities and the larger part of the population were convinced that a collapse of the dikes constituted a life-threatening risk for the entire region encapsulated in 'bathtubs'.

Ad 4 This kind of situation – low levels of threat combined with sudden onset – is difficult to imagine in the context of floods and flood management. Nevertheless, thanks to learning experiences from preceding years, in the course of the 1995 floods a number of local communities in northern France, Belgium, Germany, and the Netherlands showed a remarkable self-regulatory capacity which occasionally enabled them to suspend or even avoid evacuation. They manifested learning capacities to the effect that they were able to mitigate the effects of impending dangers which by themselves could have developed in a real menace.

Ad 5 On various occasions in 1993 authorities did not inform the population about an impending danger which by itself was not immediately life-threatening. In fact both the policymakers, the emergency services and local inhabitants tended to neglect or play down indications of steadily increasing water levels. If they would have been more alert, they could have arranged for a step by step evacuation on locations which later on would face immanent and critical situations. The underestimation of the scope and speed of the imminent threat caused delays in the evacuation process. It complicated the planning and implementation of evacuation decisions.

In the northern region of France, for instance in the Givet area, problems were caused in 1993 by groups of people who simply refused to evacuate and frustrated the efforts of rescue workers. After these sobering experiences, the authorities and the population both took a different stance in 1995, which smoothened the evacuation process. In 1993, the Belgian provinces of Namur and Limburg were taken by surprise by the sudden increase of water in the Meuse and Lesse rivers. With official warnings absent or coming in at the last minute, the population was left uninformed, and was shocked by the sudden change in threat perception that was forced upon them by the rising water. For this reason, many people were not inclined to cooperate on the basis of what they considered to be tardy warnings. Some inhabitants evacuated without a prior warning but others refused to evacuate, causing extra work (night-time evacuations) for the rescue-services. All in all, the evacuation process, which in an earlier stage could have developed smoothly, was clearly complicated by the lack of proper risk assessment and insufficient preparation.

In 1993 the southernmost part of the Dutch province of Limburg was completely surprised by the high water situation. Around Maastricht, the first evacuations had to take place in the late evening and during the night. In itself, the flood had no calamitous dimensions, but the evacuation caused more trouble than would have been possible with better preparations and a more effective use of warning time. Evacuation problems in the Netherlands in 1993 formed part of a bigger set of deficiencies associated with the lack of transnational coordination and warning agreements between the different countries existing at the time. In Belgium, no information was collected from France about the rising water levels upstream the Meuse. For the Netherlands the same applied in relation to Belgium. The contacts between the different areas were not institutionalized, and national organizations acted as if the river Meuse started and ended at their frontiers.

Ad 6 In comparison with the previous categories of floods and flood management demands, the authorities will of course prefer the more or less fortuitous combination of moderate to low threats and fully used warning time. In 1995, the majority of the cases in all four countries fell into this category. In northern France, Belgium, the Netherlands and Germany the experience of the 1993 floods made things much easier. A substantial number of authorities reaped the benefits of disaster planning initiatives they had undertaken based on the 1993 experience. The rescue services also had made adjustments in their preplanning practices and procedures. In some cities in Germany and the Netherlands rather drastic emergency measures were taken as soon as the first announcements of high water were made. The authorities were able to convince the people that a disastrous situation could occur again, and official evacuation decisions were readily accepted and complied with.

The six categories of the typology relate to a variety of evacuation patterns: from last-minute evacuations and evacuations organized during the flooding (cells 1 and 4) to well-prepared evacuations of timely briefed and well-informed populations (cells 3 and 6). But evacuations during the floods of 1993 and 1995 did not remain limited to these various patterns. For instance, there were situations in which authorities, rescue services or the population had enough time to prepare for an adequate response and for a well-planned and deliberate evacuation, but for different reasons did not evacuate or evacuated at the very last moment – at night, with again rising water causing extra trouble for the rescue services. The most common reasons were sheer situation-bound surprise, a lack of disaster planning, insufficient communication arrangements with other countries, deficient communication practices towards the public, popular denial of the coming threat, their fear to leave their homes and belongings behind, and, notably in 1995, people's gradual familiarity with the relative danger.

Furthermore, evacuation patterns varied according to the degree of perceived threat. In more dangerous or potentially dangerous situations, evacuations tended to be obligatory, raising questions about the legal status of compulsory evacuation. In Belgium, there was also a debate about the division of responsibility in case of a compulsory evacuation. In these situations the availability of sufficient time to prepare for an evacuation influenced not only the warning process, but also the extent to which the population became convinced of the severity of the coming threat. The floods of 1993 and 1995 confirm the well-known thesis that the more citizens perceive a threat to be serious, the more they are willing to evacuate.[17] For example in cases where voluntary evacuations were called for as a response to relatively small inconveniences like power failure, only small numbers of citizens followed the evacuation recommended by the authorities.

Finally, in the course of the large-scale evacuation in the Netherlands specific problems arose after the decision to evacuate had been taken. During implementation, evacuation turned out to be a highly complex and differentiated social process. The authorities had to find solutions for such special categories as businesses, hospitals and old pension homes. One of the most cumbersome elements during the whole evacuation process turned out to be the transport of tens of thousands of cattle. Furthermore, the authorities had to devise access rules and procedures in the face of the many people who wished to enter the evacuated territory, as well as those who did not want to leave their houses and, as typical stragglers, accepted the exposure to considerable risk.

A second problem related to the seemingly self-sustaining dynamics of evacuation decision making. An 'evacuation train' kept moving downstream, and stopping it created an intriguing shift of administrative rationality. Whereas during the initial stages of the Gelderland evacuations it had to be demonstrated convincingly that an evacuation was absolutely necessary, later on the burden of proof seemed to have been reversed: it now had to be shown beyond doubt that there was no reason to evacuate.

In the aftermath of the precautionary evacuation many questions were raised as to the appropriateness of the evacuation orders. As with all forms of preventive policymaking, precautionary evacuations are extraordinarily difficult to evaluate. In the Dutch case, no dikes broke, and no major damage was done. This made the evacuation look like an overreaction, or even an opportunistic act by authorities eager to dramatize flood risks to strengthen their bid for dike improvements, or exercise post-1993 disaster plans.

5 The politics of damage compensation

In addition to the damage to people and animals, floods often cause extensive financial needs. In recent years, disasters seem to lead to a decreasing number of victims on the one hand, and a growing financial damage on the other, especially in the more developed countries. The rising financial costs of floods in particular raise questions of liability for the compensation of flood damages.[18] The damage that floods cause can be distinguished into different categories. Handmer proposed the typology of flood-related damage depicted in figure 6.2 (see next page).[19]

The primary focus in this study has been on the first two categories: direct and indirect material damage. The issue of compensating these financial losses is a recurrent question. There are different options for the compensation for direct and indirect flood damage. A first option is that victims should simply absorb (part of) the damage caused by floods. A second option is private insurance.

Table 6.2 A typology of flood damages

Measurement	Type of loss	
	Direct (physical contact of property with flood water or debris)	Indirect (flood-induced disruption of economic and social activities or stress)
Material (monetary values)	1 Damage to infrastructure, buildings and contents, vehicles	2 Loss of production; clean-up costs
Immaterial (non-monetary values)	3 Death by drowning; loss of items of cultural significance and personal memorabilia	4 Inconvenience and disruption, especially to social life. Stress-induced ill-health and mortality

There are only a few countries where insurance against flooding risks is possible. Insurance companies view floods and other disasters as 'low probability, high impact' events. Floods can lead to extraordinarily high expenditures exceeding their total income gained from premiums. Furthermore, only people and organizations who live in high risk areas are willing to buy an insurance policy, increasing the risk for the insurance companies, which therefore are reluctant to offer insurance policies against flooding. The third option is to rely on social self regulation, in particular on charity organizations collecting money after a disaster and redistributing these funds to the victims. Finally, the government can choose to compensate damages. This may involve total or partial compensation, and may involve both government-supported insurance schemes and the use of non premium-related public disaster support funds.

The four country studies reveal strikingly different institutional mechanisms for handling damage compensation. We discuss them with reference to three dimensions of the problem: determining the nature and extent of damage, the public-private mix in the payment of compensation, and the implications for government responsibility.

Registering and assessing flood damages

Collecting figures about the flood damage proved to be a difficult task. In all countries, there was the problem that it is exceedingly difficult to arrive at objective, comprehensive and reliable estimations of the real costs of damages. The countries involved used different methods. The National Disaster Compensation Fund in Belgium, for instance, collects data and provides financial compensation in a strictly centralized and top-down fashion. The experiences of our French and German colleagues in trying to obtain data on overall damages illustrate how difficult it is to obtain accurate aggregate data when no specific organization is involved in collecting this kind of information. In the Netherlands, data collection was highly segmented, with information pertaining to different categories of victims (citizens, corporations, farmers) being collected by different ministries. Whatever the system chosen, essential information needed to assess damages will have to come from claims issued by victims. This presents several problems. Not everybody will submit a claim for damage compensation, and the estimates provided may be grossly inaccurate. Also, the indirect costs of flood disasters are difficult to determine. Several indirect costs emerged during the floods. For instance, the business losses sustained by firms having to close their operations and evacuate the flooded areas are sometimes hard to assess.

Indemnifications of these costs were especially difficult in countries lacking a formal system of compensation. The Netherlands are a marked example of this. Criticism and actions from farmers in 1995 led to a continuous adjustment of the settlements. The farmers demanded that all evacuation costs would be compensated for. Eventually, the government conceded to these demands almost completely. Other countries did not have these problems since indirect costs were or excluded from compensation (Belgium) or strictly circumscribed beforehand (France).

The organization of compensation

On many occasions, the damage inflicted by the floods proved to be a major burden for those affected, be they individual citizens, farms, shopkeepers or corporations. The settlements that these victims finally obtained were strongly dependent on the system of compensation that prevailed in their country of residence. Table 6.3 (see next page) elaborates the various institutional modalities that may be found. We shall discuss the four national systems in terms of this typology.

The Netherlands: ad-hoc government compensation (1a) — The issue of damage compensation attracted limited public and political attention in the Netherlands until a minor earthquake in 1992 took place in the southern part of the country. Following the 1993 and 1995 floods, the interest in the issue grew considerably.

Table 6.3 Typology of compensation systems[20]

1	**Compensation by government**
a	ad-hoc settlements
b	legal settlements
2	**Insurance systems with governmental support**
a	legal obligatory settlement
b	voluntary system with financial support from government
3	**Insurance systems without governmental support**
a	pool of insurers
b	individual insurance

But there no formal compensation arrangements for the damage caused by natural disasters were put in place. Politicians and members of parliament repeatedly stressed that government should and would compensate the damage. This resulted in intensive negotiations between the government and representatives of victim groups about the extent of coverage provided. The resulting compensation arrangements turned out to be comparatively generous and costly to the government. The reasons for this generosity were political and moral. The political considerations were directly related to the fact that the floods occurred between national elections. As to the moral concerns involved, the over-all notion was that the authorities had neglected the maintenance of the river dikes and by this omission helped to maintain infrastructural vulnerabilities that allowed for considerable flooding in 1993 and necessitated the evacuations that produced much of the total damage in 1995.

The generous attitude on the part of the government had some negative consequences. The settlements were open-ended and had no upper limits. Consequently, they imposed a considerable budgetary burden. Furthermore, it could be expected that in the absence of a more formalized compensation scheme, in future disasters people will turn to the government using prior flood damage payments as a precedent. The problems of this kind of analogical reasoning came to the fore following the 1995 flood. Victims claimed compensation from the government akin to 1993 levels. However, from a technical insurance standpoint, it was not at all evident that the two cases should be treated identically, since there are major qualitative differences between the 'wet' damage caused by physical contact with the water and 'dry' damage such as evacuation costs in areas where no flooding actually took place. The eventual outcome was the result of ad hoc pulling and hauling between stakeholders and government agencies instead of a transparent, predictable and widely agreed upon insurance logic.

Belgium: law-based compensation (1b) — Compensation in Belgium is administered by the National Disaster Compensation Fund. A special law provides compensation for exceptional natural events of unforeseeable calamities which have caused major damage, notably earthquakes or tremors, flooding, hurricanes or other high-force winds. The event should be officially designated as a disaster before compensation is paid. This recognition should take the form of a Royal Decree issued by the Minister of the Interior in consultation with the Council of Ministers. The victims of the floodings can then submit their claims for up to six months after the declaration has been issued. Claims are judged by provincial governors, and after their agreement payment is managed by the Fund.

During the 1993 and 1995 floods, there was little doubt that these events would be regarded as disasters eligible for compensation by the National Disaster Compensation Fund. Eventually, 3.632 claims were made in 1993 and 2.400 claims in 1995. Where the main weakness of the Dutch system was on the input side, e.g., the unstructured bargaining about claims and compensation levels, the Belgian system suffered from severe implementation problems. The processing of claims and the payment of compensation progressed slowly and arduously. The Fund was understaffed, its forms were user-unfriendly, its payment procedures were cumbersome, and above all no time limit was set for the provincial governors to decide upon the claims. Ironically at the time of the 1995 floods compensations for the 1993 floods were yet to be paid. The federal government responded to these problems by establishing a task force to look for ways to speed up the process.

France: government-supported obligatory insurance (2a) — Damage compensation in France is organized by a by law in which premiums are set by ministerial order. The insurance is based on a basic policy insurance, usually the fire insurance. Payment under this system can only begin after an official disaster declaration has been issued. This declaration is made after the departmental prefect has put together a file determining the nature and intensity of the catastrophe as well as a list of the affected communes. Eventually, the Ministry of the Interior, the Ministry of Economy and Finance and the Ministry of the Budget elevate the declaration to the status of an inter-ministerial order and publish it in the *Journal Officiel*.

The clear advantage of the French system is that fixed criteria are set by law. Moreover, eighty to ninety percent of the indemnifications are paid within the legal time period (3 months) and conflicts about compensation levels are rare. The existence of this procedure seems to minimize the uncertainty of victims and stricken communities in the aftermath of the floods, even though its defacto coverage was much lower than the eventual outcome in the Netherlands. This was probably enhanced by the availability of additional sources of financial compensation. The experiences in the Vaucluse and the Ardennes show that large amounts of money that may become available through spontaneous gifts and alternative funds made available by local organizations, charity associations and the European Union. Nevertheless, the system used in France has some negative side-effects too. Firstly, the insurance companies are said to profit too much from the system. The possibility of reinsuring the risks with the Central Fund for Reinsurance is hardly used, because the insurers are capable of insuring the disaster-related risks themselves. Secondly, the insurance firms are imposing more restrictions on areas with a relatively high risk. This can lead to situations where insurance policies are actually canceled. This happened in the commune of Bédarrides (Vaucluse) where an insurance group canceled the policies because of the high frequency of floods.

Germany: private insurance markets (3b) — Except for the state of Baden Würtemberg, no public flood insurance exists in Germany. The burden of losses falls primarily on home and business owners themselves, who have to operate on a completely market driven insurance scheme. The different *Länder* do grant tax depreciation provisions to the victims to write off the costs of the flood damage, yet they pay little or no specific compensation and grant only general tax deductions. This system may come under increasing pressure in the face of the widening gap between the political and public perception about the causes of riverine floods and the management of flood damage by the authorities. If the public comes to perceive government negligence to be a major contributing cause of the floods, political pressures will mount on policymakers to admit liability and provide some form of public compensation to flood victims.

Every country thus has its own way of organizing the compensation of flood-related damage. A key policy question with respect to these various schemes and arrangements is how and to what extent the government contributes to damage compensation. Compensations in the Netherlands and Belgium depend completely on governmental resources, while insurance companies play a more prominent role in France. In Germany, the contributions on the part of the government are minimal. Insurance is possible with private companies offering individual policies against fairly high costs. The question of governmental responsibility in the case of disasters is likely to become ever more prominent. People increasingly expect protection from government against such risks as flooding. In contemporary Western Europe, floods – like many other spells of nature such as earthquakes or forest fires – are at best only partially seen as natural disasters. For a large part, they are becoming framed as policy disasters, e.g., failures of prevention and mitigation. Acts of God or misfortune are hardly seen as legitimate policy excuses anymore when a modern community is rocked by unpleasant or even traumatic events.[21] Allegations of government negligence have become part and parcel of the politics of blaming following major disasters that – at least in the public view – could and should have been prevented. For example, within 48 hours after a 1996 flash flood destroyed a Spanish campsite media and political spotlights were on the European Commission that was said to have been less than vigorous in producing and enforcing legislation concerning campsite safety. Public irritation, even outrage, at perceived government indifference or incompetence may lead to a serious loss of legitimacy for political and bureaucratic elites. And when citizens expect the authorities to protect them, they will call upon the government for legitimate compensation whenever such protection is seen to have failed.

Long-term planning efforts aiming at flood and disaster management should therefore pay appropriate attention to damage compensation and insurance arrangements. Governments should be aware though that total dependence of such schemes on public resources might have tremendous financial consequences. Public-private cooperation involving both governmental investments and an active role of insurers or other associations can result into a viable spreading of risks. However, the intense post-disaster debates about liability in Germany are just one example of the fact that times have passed when governments could stay aloof from any responsibility for the material consequences of disasters which one way or the other raise questions about government pre-disaster policies.

6 Organizational learning

There are considerable differences between the four countries as to the most remarkable features of the floods examined: the mass evacuations in the Netherlands; the flash floods in France; the specific political and administrative situation in Belgium, with the fairly new federation of Wallonian and Flemish communities; the typical characteristics of urban flood management in Germany. From this angle, the 1993-1995 events can be seen as an unusual learning opportunity for the authorities and agencies involved.[22]

Learning processes played an important role in the domain of risk communication. Prior experience proved crucial in explaining the rather marked contrast between the 1993 and 1995 warning process in riverine flood communities. In 1993 the floods, notably their scope and intensity, came as a big surprise to the German, Belgian and Dutch authorities and emergency services. This made for serious delays in the issuing of adequate warnings to the public. In 1995, the key officials and agencies were much more alert. Their cognitive repertoire had been broadened by the 1993 experience, changing their basic attitude from 'it will probably blow over' to 'this might turn out to be serious'. This made for a more activist approach, and a decreased reluctance to issue warnings that later may turn out to have been exaggerated. In this case, therefore, the 'lessons of history' suggested by the problems encountered during the 1993 episodes proved beneficial in sensitizing authorities and citizens to the disaster potential of floods.

In fact, all four countries showed an upward learning curve in other aspects of disaster management as well:

- In France, the flood in Vaison-la-Romaine in 1992 worked to focus the local authorities' attention to campsites.
- In Belgium, in 1995 coordination activities were more effective than thirteen months before. The rescue services handled the 1995 floods in a much more routinized fashion.
- In the Netherlands, municipalities in the provinces of Limburg and Gelderland made special emergency plans after the 1993 floods. These planning processes contributed significantly to successful disaster management in 1995. Last-minute precautionary measures taken in 1995 on the basis of the 1993 experience prevented the infliction of much damage.
- In Germany, the 1993 experience led to improvements in the organization of disaster management, amounting to a more decentralized structure. The operational centers assumed more responsibilities than before.

Different dimensions of learning can be distinguished in the European flood management experience. Firstly, *diachronical learning* pertains to learning across different flooding episodes. In 1995, authorities and emergency services were confronted with a second flood within a relatively short period of time. Between the 1993 and 1995 flood, they were able to improve flood management in several ways: preventive measures and last-minute efforts regarding emergency dike provisions; organizational adjustments, for instance assistance on the part of the military; improvements in communications; the development of specific emergency plans.

Secondly, the *focal point* of the lessons learned may differ. Learning can take place in each of the main phases of the disaster management process: prevention, preparation (disaster planning and warning), response (assistance) and recovery (material and immaterial recovery). The 1993-1995 learning process concentrated primarily on the state of preparedness and response. This is not surprising. The short timespan between the two episodes was not enough to make and implement major political decisions regarding the elaborate infrastructural improvements that would be needed along the banks of many rivers to increase the effectiveness of prevention. When it came to preparation and response, people facing a similar situation as in 1993 actually tested the post-1993 measures in a real-life situation.

The cumulative disaster experience of the 1993 and 1995 floods did prove to be a political window of opportunity for the proponents of improved prevention through dike reconstruction and waterways management.[23] In Holland, for example, the conservationists and environmentalists that had long succeeded in stalling dike improvement schemes were quickly cast to be scapegoats following the 1995 mass evacuations. In some communities, they were even physically harassed. Politically, they lost momentum, and emergency legislation enabling dike improvements was pushed through parliament. In the longer run, a more balanced approach to combining reconstruction and conservation is likely to emerge.

Thirdly, the 1993 and 1995 floods in several countries in Western Europe were occasions for *spatial learning*. The floods enabled the exchange of experiences across communities, regions, and indeed states. This type of learning was limited at best. The transboundary dissemination and diffusion of knowledge presupposed by the spatial learning concept did not really materialize between 1993 and 1995.

More generally, productive learning from disaster experiences is by no means guaranteed. For example, 'near-miss' experiences with flooding prior to 1993 were a key factor explaining the passive if not complacent attitude taken by many in Germany, Belgium and Holland when the water first started rising in 1993. In 1995, in contrast some authorities tended to ask for more assistance than was strictly necessary. Similar practices could be traced with citizens who had learned their lessons with regard to adopting a firm stance in the litigation process. Quite some people refused to evacuate because two years before their evacuation had turned out to be unnecessary.

The fixation on one single period of time (1993-1995) or one single and specific category of events (floods) could easily lead to solutions overlooking the unique character of disasters yet to come. Studies of crisis management in other domains also show how recent 'success' experiences may desensitize decision makers and lock them into a policy formula that is applied uncritically to the next crisis and is rigidly maintained even in the face of repeated indications of its ineffectiveness in the changed circumstances.[24]

This suggests that active steps may need to be taken to prevent the generally positive experiences with risk communication and warning in 1995 from providing authorities with a false sense of self-confidence and security. The scenario of the next floods is quite likely to be rather different from the 1995 script, especially given the major infrastructural works that have been undertaken at various points along the rivers, and that will change the distribution of risks in ways that are not yet fully understood.

Clearly, learning must have a long-term orientation and must be institutionalized in the communities and governments involved. It should be enhanced by plans, budgets and routines, and not just be dependent upon the commitment of individual community leaders and agency officials – who may move, retire, be promoted, or lose interest as their memory of the last flood begins to fade. Institutional learning from floods means learning from many experiences at different places over a considerable span of time. This study presents only parts of the problem of flood management. Steady attention and the willingness to learn from experiences provide the best opportunities to cope with floods in general. Dissemination of knowledge between the countries of Europe is necessary to get a full picture of floods and flood management as well as of disasters and disaster management in general.

Notes

1 D.J. Parker, J. Handmer, *The role of unofficial flood warning systems*, Unpublished paper, Flood Hazard Research Centre, Middlesex University 1996.

2 Cf. J.M. Weller, D. Wenger, *Disaster subcultures: The cultural residues of community disasters*, Paper presented at the Annual Meeting of the North Central Sociological Association, Cincinnati 1973; D.S. Mileti, T.E. Drabek, J.E. Haas, *Human systems in extreme environments: A sociological perspective*, Boulder: Institute of Behavioral Science, University of Colorado 1975.

3 See D. Parker, M. Fordham, J-P Torterotot, Real-time hazard management: Flood forecasting, warning and response, in: E. Penning-Rowsell and M. Fordham (eds), *Floods across Europe: Flood hazard assessment, modelling and management*, London: Middlesex University Press 1994, pp. 135-166.

4 See Parker and Handmer, *op. cit.*

5 This is a classical problem of disaster warning, see E.L. Quarantelli, V. Taylor, *Some views on the warning problem in disasters as suggested by sociological research*, Working Paper 45, Ohio State University: Disaster Research Centre 1977, p. 4.

6 The problems of intergovernmental relations in emergency management have received ample attention in US didsaster scholarship. See W.J. Petak (ed.), Special issue on emergency management, *Public Administration Review*, 1985-1; L. Comfort (ed.), *Managing disaster*, Durham: Duke University Press 1988; R.M. Stratton, *Disaster relief: The politics of intergovernmental relations*, Lanham: University Press of America 1989; S.K. Schneider, *Flirting with disaster: Public management in crisis situations*, New York: M.E. Sharpe 1995, pp. 28-39. An informative French study of the same problem is C. Gilbert (dir.), *La catastrophe, l'élu et le préfet*, Grenoble: Presses Universitaires de Grenoble, 1990. A German case study is R. Czada, Politics and administration during a 'nuclear-political' crisis, *Contemporary Crises*, 14, 1990, 285-312. The Dutch context is described in U. Rosenthal, Disaster management in the Netherlands: Planning for real events, in: Comfort (ed.), *op. cit.*, pp. 274-296.

7 See P. 't Hart, U. Rosenthal, A. Kouzmin, 'Crisis decision making: The centralization thesis revisited,' *Administration and Society*, 25, 1993, 12-45.

8 R.R. Dynes, *Organized behavior in disaster*, Lexington: Heath 1970 provides a classic typology of organizations that become active during disasters.

9 See P. 't Hart, B. Pijnenburg, The Heizel stadium tragedy, in: Rosenthal, Charles and 't Hart (eds), *op. cit.*, pp. 197-224; U. Rosenthal, P. 't Hart, A. Kouzmin, The bureau-politics of crisis management, *Public Administration*, 69, 1991, 211-233.

10 See, for example, A.H. Barton, *Communities in disaster*, New York: Doubleday 1969; J.B. Perry, M.D. Pugh, *Collective behavior: Response to social stress*, New York: West Publishing Company 1978; E.L. Quarantelli (ed.), *Disasters: Theory and research*, Beverly Hills: Sage 1978; R. Gist, B. Lublin (eds), *Psychosocial aspects of disaster*, New York: Wiley 1989; G. Kreps, Foundations and principles for emergency planning and management, in: D. Parker, J. Handmer (eds), *Hazard management and emergency planning: Perspectives on Britain*, London: James and James 1992, pp. 159-174.

11 L.M. Walters, L. Wilkins, T. Waters (eds), *Bad tidings: Communication and catastrophe*, Hillsdale: Lawrence Erlbaum 1989; P. Lagadec, *La gestion des crises: Outils de réflexion à l'usage des décideurs*, Paris: Mc Graw-Hill 1991, pp. 111-141.

12 U. Rosenthal et al., *Complexity in urban crisis management: Amsterdam's response to the Bijlmer air disaster*, London: James and James 1994. Incidentally, tourists in the literal sense have also become the object of attention in disaster research. They are a vulnerable category due to their unfamiliarity with local conditions and administrative structures, and their lack of an informal social support structure of family and friends. See T.E. Drabek, Understanding tourists during disaster, *Australian Journal of Emergency Management*, 11, 1996/97, 14-19.

13 P. 't Hart, Symbols, rituals and power: The lost dimension of crisis management, *Journal of Contingencies and Crisis Management*, 1, 1993, 27-41.

14 See, for example, K.T. Erikson, *Everything in its path*, New York: Simon and Schuster 1976.

15 This typology is developed in Crisis Onderzoek Team, *Evacuaties bij hoog water: Zelfredzaamheid en overheidszorg*, Leiden/Rotterdam: Crisis Onderzoek Team 1995.

16 Because of a lack of reliable information about evacuations in Germany, these are not included in the table.

17 Drabek, *op. cit.*, pp. 71-99.

18 C. Wells, *Negotiating tragedy: Law and disasters*, London: Sweet and Maxwell 1995.

19 J.W. Handmer, The evolving flood threat, in: *Australian disaster and emergency management: How good?*, Mt Macedon: Australian Counter Disaster College 1991, p. 181.

20 Based on: K.P. Barnhoorn, Overstroming en aardbeving, *Beursbengel*, 637, augustus 1995.

21 M. Bovens, P. 't Hart, *Understanding policy fiascoes*, New Brunswick: Transaction Books 1996.

22 There is an emerging literature on disasters and organizational learning, for example M.J. van Duin, *Van rampen leren*, Den Haag: De Haagse Drukkerij 1992 and L.K. Comfort, Risk and resilience: Interorganizational learning following the Northridge earthquake of 17 January 1994, *Journal of Contingencies and Crisis Management*, 2, 1994, 157-170; P. Lagadec, Learning processes for crisis management in complex emergencies, *Journal of Contingencies and Crisis Management*, 5, 1997, 24-31.

23 Compare R.A. Boin, M.H.P. Otten, Beyond the crisis window for reform: Some ramifications for implementation, *Journal of Contingencies and Crisis Management*, 4, 1996, 149-161.

24 For example, U. Rosenthal, P. 't Hart, Managing terrorism: The South Moluccan hostage takings in the Netherlands, 1975 and 1977, in: Rosenthal, Charles, 't Hart (eds), *op. cit.*; E. Stern, B. Sundelius, Sweden's twin monetary crises of 1992: Rigidity and learning in crisis decision making, *Journal of Contingencies and Crisis Management*, 5, 1997, 32-48.

Appendix 1

SOURCES

France

SOURCES AND MATERIALS USED

- Interviews in the two departments concerned.
- Systematic analysis of the national and local press for the periods and locations studied.
- Viewing of audio-visual documentation.
- Analysis of non-official, informal reports and notes, laws, decrees, orders, official reports, internal memos and any documents provided by the institutions and services concerned.

OTHER DOCUMENTS

- BIP, Bulletin d'Information du Personnel de la mairie de Charleville-Mézières, nr. 22, avril 1995.
- Bollène, L'information municipale en direct, 'Spécial inondations. 1er octobre 94: un an après', édité par la ville de Bollène, supplément au nr. 7, 1994.
- Bollène, L'information municipale en direct, 'Spécial : 1993...94-95, le rebond', septembre-octobre 1995.
- Bulletin d'information municipale, Ville de Charleville-Mézières, nr. 68, novembre 1994.
- Bulletin d'information municipale, Ville de Charleville-Mézières, nr. 70, mars 1995.
- Caille, Frédéric, *Institutions au péril de l'eau: Approche du système d'acteurs de la politique de prévention du risque inondation*, Rapport, Commissariat Général du Plan-Instance d'Evaluation de la Politique Publique de Prévention des Risques Naturels – Futuribles, septembre 1994.
- Canepa, Daniel, 'La sécurité civile face aux inondations', *Les Cahiers de la Sécurité Intérieure*, nr. 16, 2ème trimestre 1994, 113-120.
- Cans, Chantal dir., 'L'indemnisation des victimes de catastrophes naturelles: dossier', *Sécurité, Revue de Préventique*, nr. 10, janvier-février 1994.

- Conseil Général des Ardennes, Direction Général des services départementaux, Direction des routes et des infrastructures, *Rapport relatif aux dégâts occasionnés par les inondations et intempéries de février 1995 (infrastructures routières)*, 1995.
- Coupy, Jacques, *Inondations tranquilles et torrentielles*, rapport RESPOM, Conseil Général des Ponts et Chaussées, septembre 1994.
- De Bagnolo, Vittoria, 'Inondations: prévenir plutôt que guérir', *L'Assurance Française*, nr. 705, janvier 1995.
- Direction Départementale des Services d'Incendie et de Secours, *Inondation Valréas-Bollène, 30 septembre-8 octobre 1993*, Sapeurs-pompiers du Vaucluse.
- Direction Départementale des Services d'Incendie et de de Secours, *Inondation septembre-octobre 1993, Conclusion, récapitulatif: Dégâts matériels, bilan financier*, Sapeurs-pompiers du Vaucluse.
- Dye, Vincent, *La sécurité civile en France*, Paris, PUF, mars 1995.
- Direction de la Sécurité Civile, Sous-direction des opérations de secours, Bureau de la coordination opérationnelle, *Retour d'expérience 1992-1994*, rapport, 1994.
- Gilbert, Claude, 'Le traitement des catastrophes: entre assurance et solidarité', rapport pour la table ronde *'Affronter les catastrophes': Entretiens de l'Assurance*, FFSA, Paris, 13 décembre 1993.
- Instance d'Evaluation de la Politique Publique de Prévention des Risques Naturels, *Rapport d'étape*, Commissariat général du Plan, 24 mars 1995.
- Lambert-Faivre, Yvonne, 'L'indemnisation des victimes de catastrophes naturelles', Communication au Colloque, *'La crise et le droit de la sécurité civile'*, Avignon, 14-15 mai 1993.
- Ministère de l'Intérieur – Inspection générale de l'administration, *Le préfet et la crise: Charte du comportement*, juillet 1994.
- Noyelle, Jean, *Prévention et gestion des phénomènes hydrologiques exceptionnels*, note, DPPR, ministère de l'Environnement, novembre 1993.
- Préfecture des Ardennes, *Inondation de janvier et février 1995: Chronologie des événements survenus dans le département des Ardennes, période du lundi 23 janvier 1995 au mercredi 8 février 1995*, 1995.
- Préfecture de la Zone de Défense Est, Etat-Major de Zone-CIRCOSC, *Les inondations en Zone de Défense Est du 22 janvier au 8 février 1995*, mars 1995.
- *Rapport de la Commission d'enquête sur les causes des inondations et les moyens d'y remédier*, rapport nr. 1641, Assemblée Nationale, Les documents d'information de l'Assemblée Nationale, tome 1 et 2, 1994.
- *Rapport sur l'évaluation des dégâts causés par les inondations de l'été et de l'automne 1993 dans le sud de la France*, présenté par Gilles Sanson et Tristan Florenne, Ministère de l'Intérieur et de l'Aménagement du Territoire, juillet 1994.
- *Rapport sur l'évaluation des dégâts causés en France par les inondations de l'hiver 1993-1994*, présenté par Gilles Sanson et Tristan Florenne, Ministère de l'Intérieur et de l'Aménagement du Territoire, juillet 1994.

- *Le Tour de Ville: Revue d'information de l'Association Valréas Communication*, nr. 8, 4ème trimestre 1994 et nr. 10, 4ème trimestre 1994.

Belgium

SOURCES AND MATERIALS USED

- Analysis of reports, minutes of meetings, updates, expense statements, evaluation reports of various organisations (in)directly involved in the emergency operations in Namur and Limburg.
- Analysis of the news bulletins on the floods transmitted by the various media.
- Supplementary questionnaire sent to the heads of the emergency service organisations concerned in the province of Namur.
- Study of documents at the Government Coordination and Crisis Centre.

OTHER DOCUMENTS

- Clymans A., De Belgische noodplanning: de nucleaire toepassing, *Annalen van de Belgische Vereniging voor Stralingsbescherming*, 18, 1993, 249-273.
- Contrat de Riviere Haute Meuse, *Vallee de la Haute Meuse Belge, Inondations: Que faire?*, Namur 1995.
- De rivier is verdronken. Een bericht uit de Maasvallei, *Knack*, 1 februari 1995, 23-25.
- Dienst Hydrologisch Onderzoek, *Summiere nota met voorlopige gegevens over de hoge waterstanden en debieten in de belangrijkste, niet aan het getij onderhevige waterlopen van het Vlaamse Gewest*, Dienst Hydrologisch Onderzoek, Brussel 1994.
- Heylen, J., *De dienst hydrologisch onderzoek van het Vlaams Departement van Leefmilieu en Infrastructuur en de opvolging in reële tijd en de voorspelling van hoge waterstanden in perioden van hoge waterstanden*, Dienst Hydrologisch Onderzoek, Brussel 1995.
- Hoger Instituut voor de Noodplanning, *Filosofie en doelstellingen van de Belgische noodplanning*, Florival 1992.
- Hoger Instituut voor de Noodplanning, *Rol van de media*, Florival 1992.
- Hoger Instituut voor de Noodplanning, *Voorlichting en communicatie tijdens crisissituatie*, Florival 1994.
- Hoger Instituut voor de Noodplanning, *Organisatie en werking van een commando- of coördinatiepost*, Florival 1994.
- Hysman, H., Maasoverstroming januari-februari 1995, *De Belgische Brandweerman*, 1995, nr. 1, 12-17.
- Koninklijk Meteorologisch Instituut, De overstromingen van januari 1995 ... 13 maanden later, in: *Ring*, 1995, 9-11.

- Koninklijk Meteorologisch Instituut, *Klimatologisch overzicht van december 1993, januari 1994, januari 1995 en februari 1995*, KMI-IRM, Dienst Klimatologisch Onderzoek.
- Le béton et le coquelicot; Inondations 1993, *Le vif/L'Express*, 31 décembre 1993, 6-9.
- Meers, H., 'Maaseik kan overstromen! Ligt 2 meter onder het waterpeil! De burgemeester is de wanhoop nabij!': Dagboek van een crisismanager in Maasmechelen, *Alert*, 1994, nr. 2, 20-22.
- Meers, H., De crisis bezworen: in het crisiscentrum van Belgisch Limburg, *Alert*, 1994, nr. 2, 18-19.
- Meers, H., *Het informatiebeleid naar aanleiding van de overstromingen van de Maas van december 1993 in Limburg*, Florival, Hoger Instituut voor de Noodplanning, 1994.
- Ministere Wallon de l'Equipement et des Transports, *Sethy - Service d'Etudes Hydrologiques*, Bruxelles, Direction Générale des voies hydrauliques.
- Ministerie van Binnenlandse Zaken, *Informatiedagen noodplanning*, Koninklijke School van de Civiele Bescherming, Florival 1990.
- Province de Namur, *Plan General de Secours de la Province de Namur*, Namur 1988.
- Provinciebestuur Limburg, *Algemeen Noodplan*, Hasselt, 1993.
- Provinciebestuur Limburg, *Overstroming Limburgse Maasvallei einde december 1993; Eindevaluatieverslag over de hulpverlening*, Hasselt 1994.
- Vandiepenbeeck M. et al., A propos des récentes inondations en Belgique, *Ciel et Terre*, Institut Royal Météorologique de Belgique, 111, 1995, 47-54.

The Netherlands

SOURCES AND MATERIALS USED

- First hand observation in various provincial, regional and municipal crisis centres during both floods.
- Interviews with key decision makers at various levels of government and in a variety of counter-disaster agencies. For the damage assessment and liability part, interviews with business representatives in all affected areas following both floods.
- Analysis of a comprehensive collection of public and confidential records of the crisis centres, evaluation reports, minutes of local council debates, other investigation reports.
- Content analysis of media coverage.
- Phone survey conducted among the population in the various stricken areas both in 1993 and 1995.

OTHER DOCUMENTS

- Aguirre, B.E., Evacuation in Cancun during hurricane Gilbert, *International Journal of Mass Emergencies and Disasters*, 1, 1991, 31-45.
- Baker, E.J., Hurricane Evacuation behavior, *International Journal of Mass Emergencies and Disasters*, 1, 1991, 287-310.
- Barnhoorn, K.P., Overstroming en aardbeving (deel 1 en 2), *Beursbengel*, 637, augustus 1995.
- Barton, A.H., *Communities in disaster: a sociological analysis of collective stress situations*, New York 1969.
- Burton, I., P. Victor and A. Whyte, *The Mississauga evacuation: Final report*, The Institute for Environmental studies, Ontario 1981.
- Carter, T.M., S. Kendall and J.P. Clark, Household response to warnings, *International Journal of Mass Emergencies and Disasters*, 1, 1983, 95-104.
- Charles, M.T., *The great flood of '82*, School of Public and Environmental Affairs, Indiana University, Fort Wayne, Indiana 1985.
- Commissie Boertien II, *Onderzoek watersnood Maas*, Waterloopkundig Laboratorium in opdracht van Ministerie van Verkeer en Waterstaat, december 1994.
- Crisis Onderzoek Team, *Kritieke momenten: studies over beslissen in moeilijke omstandigheden*, Arnhem: Gouda Quint 1990.
- Crisis Research Center, *Complexity in Urban Crisis Management: Amsterdam's Response to the Bijlmer Air Disaster*, London: James & James 1994.
- Cutter, S. and K. Barnes, Evacuation behavior and Three Miles Island, *Disasters*, 2, 1982, 116-124.
- Drabek, T.E. and J.S. Stephenson, When disaster strikes, *Journal of Applied Social Psychology*, 1, 1971, 187-203.
- Drabek, T.E., *Human system responses to disaster: An inventory of sociological findings*, New York: Springer 1986.
- Duclos, P. and L. Sanderson, et al., Community evacuation following a chlorine release, Mississippi, *Disasters*, 11, 1987, 286-289.
- Duin, M.J. van, *Van rampen leren*, Den Haag 1992.
- Duin, M.J. van, Rampenbestrijding, *Bestuur*, 7, 1989, 211-215.
- Dynes, R., *Organized behavior in disaster*, Lexington: Heath 1970.
- Ellemers, J.E., *De Februari-ramp: Sociologie van een samenleving in nood*, Assen: Van Gorcum 1956.
- Erikson, K., *A new species of trouble: Explorations in disaster, trauma and community*, New York: Basic Books 1994.
- Groenendijk, J.N.A. (ed.), *Public relations en voorlichting cases: Handboek voor interne en externe communicatie*, Alphen aan den Rijn: Samsom 1995/2.
- Hart, P. 't, U. Rosenthal and A. Kouzmin, Crisis decision making: The centralization thesis revisited, *Administration and Society*, 25, 1993, 12-45.
- Hewitt, K. (ed.), *Interpretations of calamity*, Boston 1983.

- Jong, M.A.D.W. de, H.R.B.M. Kummeling & M.C. Burkens, *Het gebruik van gemeentelijke noodbevoegdheden*, Zwolle: Tjeenk Willink 1994.
- Lagadec, P., *States of emergency: Technological failures and social destabilization*, Paris 1988.
- Lagadec, P., *Preventing chaos in a crisis: Strategies for prevention, control and damage limitation*, London: McGraw-Hill 1993.
- Leenders, P., Hoog Water: rampbestrijding volgens plan, *Openbaar Bestuur*, 5, 1995, 22-25.
- Mileti, D.S., T.E. Drabek and J.E. Haas, *Human systems in extreme environments: A sociological perspective*, Institute of behavioral Science, The University of Colorado, Boulder 1975.
- Ministerie van Binnenlandse Zaken, *Handleiding Rampenbestrijding*, Den Haag 1990.
- NIBRA, *Watersnood 1995: een terugblik voor de toekomst*, Arnhem: NIBRA 1995.
- Perrow, C., *Normal accidents: Living with high risk technologies*, New York: Basic Books 1984.
- Perry, R.W., *The social psychology of civil defense*, Lexington: Heath 1982.
- Perry, R.W., *Comprehensive emergency management: evacuating threatened populations*, Greenwich: Quorum 1985.
- Perry, R.W. and J.M. Nigg, Emergency Management Strategies for Communicating Hazard Information, *Public Administration Review*, 1985, 72-76.
- Quarantelli, E.L., Some basic misconceptions, *Social Problems*, 8, 1960, 68-79.
- Quarantelli, E.L., General and particular observations on sheltering and housing in American disasters, *Disasters*, 4, 1982, 277-280.
- Quarantelli, E.L., *Evacuation behavior and problems: Findings and implications from the research literature*, Delaware: Disaster Research Center 1984.
- Quarantelli, E.L. and V. Taylor, *Some views on the warning problem in disasters as suggested by sociological approach*, Columbus: Disaster Research Center, 45, 1977.
- Raphael, B., *When disaster strikes: A handbook for the caring professions*, London 1986.
- Research voor Beleid, *Voorlichting bij rampen: Een inventariserende literatuurstudie*, Leiden 1987.
- Rosenthal, U., Studies in the Holland flood disaster 1953, an essay on the proto sociology of disaster, *International Journal of Mass Emergencies and Disasters*, 3, 1988, 233-251.
- Rosenthal, U., *Rampen, rellen, gijzelingen: crisisbesluitvorming in Nederland*, Amsterdam/Dieren: De Bataafse Leeuw 1984.
- Rosenthal, U., M.T. Charles and P. 't Hart (eds), *Coping with crises: The management of disasters, riots and terrorism*, Springfield: Charles Thomas 1989.
- Scanlon, T.J., Toxic chemicals and emergency management: The evacuation of Mississauga, Ontario, Canada, in: U. Rosenthal, M.T. Charles and P. 't Hart

(eds), *Coping with crises: The management of disasters, riots and terrorism*, Springfield: Charles Thomas 1989, pp. 303-322.
- Stallings, R.A., Ending Evacuations, *International Journal of Mass Emergencies and Disasters*, 2, 1991, 183-200.
- Vogt, D. and B. Muller, Issues in nursing home evacuations, *International Journal of Mass Emergencies and Disasters*, 2, 1991, 247-265.
- Wildavsky, A., *Searching for safety*, New Brunswick: Transaction 1988.
- Zelinsky, W. and L.A. Kosinsky, *The emergency evacuations of cities: A cross-national historical and geographical study*, Maryland 1991.

Germany

SOURCES AND MATERIALS USED

- Interviews with administrators at the different levels of administration and decision makers at the communal level, and with representatives of the different disaster relief services, re-insurances and hydrological and meteorological research institutes.
- Analysis of public evaluation and investigation reports.
- Evaluation of reports of the major re-insurances.
- Content analysis of the media coverage of the flood events.

OTHER DOCUMENTS

- lbers-Hain, H., Hochwasser in Trier, *Brandschutz, Deutsche Feuerwehrzeitung*, 48, 1994, 884-889, Stuttgart.
- Bayerische Rückversicherung, *The 'Christmas floods' in Germany 1993/94*, 16, München 1994.
- Bayerische Rückversicherung, *...13 Monate später*, Das Hochwasser vom Januar 1995, Sonderdruck 17, München 1995.
- *Bericht zur Zivilen Verteidigung*, Dem Parlament vorgelegt von der Bundesregierung im März 1994.
- Berkenhoff, K., Hochwassereinsatz des THW in Köln, *Brandschutz, Deutsche Feuerwehrzeitung*, 50, 1996, 145-146, Stuttgart.
- Björnsen, G., Systematik der Hochwasser-schadensminimierung, in: *Ratgeber, Mit dem Hochwasser leben*, 1995, pp. 30-35.
- Bühl, W.L., *Krisentheorien: Politik, Wirtschaft und Gesellschaft im Übergang*, Darmstadt 1984.
- Bulletins of the Ministry of the Interior of Baden-Württemberg concerning the flood 1995.
- Bulletins of the Ministry of the Interior of North Rhine-Westphalia concerning the flood 1995.

- Bulletins of the Ministry of the Interior of Rhineland-Palatinate concerning the flood 1995.
- Bundesanstalt für Gewässerkunde (BfG), *Das Hochwasser 1993/94 im Rheingebiet*, Koblenz 1994.
- Bundesanstalt für Gewässerkunde (BfG), *Hochwasser im Rheingebiet: Ursachen und Entwicklungen sowie die Ereignisse 1994 und 1995 im vieljährigen Vergleich*, Koblenz 1995.
- Bundesanstalt für Gewässerkunde (BfG), *Das Januarhochwasser 1995*, Koblenz 1995.
- Clausen, L., W.R. Dombrowsky, *Einführung in die Soziologie der Katastrophen*, Bonn 1983.
- Coellen, B., H. Schuh (o.A.): Verhalten von Menschen bei Unglücksfällen und Katastrophen, in: *'Impulse'-Fachinformation (DRK) für Führungskräfte*, Heft 2, Bonn.
- Crozier, M., E. Friedberg, *Die Zwänge kollektiven Handelns: Über Macht und Organisation*, Frankfurt am Main. 1993.
- Dienel, P.C., *Die Planungszelle: Der Bürger plant seine Umwelt. Eine Alternative zur Establishment-Demokratie*, 3. Aufl., Opladen 1992.
- Dolle, H.-E., Saarbrücken 1993: Hochwasserkatastrophe, *Brandschutz, Deutsche Feuerwehrzeitung*, 48, 1994, 878-883, Stuttgart.
- Dombrowsky, W.R., Der Mensch in der Katastrophe, in: *'Impulse'-Fachinformation (DRK) für Führungskräfte*, 1987, Heft 1, 3/1987, Bonn.
- Dombrowsky, W.R., *Krisenkommunikation. Problemstand, Fallstudien und Empfehlungen*, Kiel 1991.
- Dombrowsky, W.R., *Bürgerkonzeptionierter Zivil- und Katastrophenschutz. Das Konzept einer Planungszelle Zivil- und Katastrophenschutz*, Bonn 1992.
- Dombrowsky, W.R., Katastrophentourismus - vom Gaffer zum Helfer?, in: *Der Polizeipräsident in Berlin*, Katastrophenkommission: Seminar über Großschadensereignisse – 4. Katastrophenseminar, Berlin 1994.
- Engel, H., Sind Hochwasser beherrschbar? in: *Ratgeber. Mit dem Hochwasser leben*, 1995, pp. 6-18.
- Feyrer, J., Einsatz überörtlicher Kräfte, *Brandschutz, Deutsche Feuerwehrzeitung*, 48, 1994, 854-857.
- Feyrer, J., Am Rande vermerkt ..., *Brandschutz, Deutsche Feuerwehrzeitung*, 48, 1994, 870-871.
- Feyrer, J., Wenn die Feuerwache zur Sandburg wird, *Brandschutz, Deutsche Feuerwehrzeitung*, 50, 1996, 139-144.
- Fischer, H., *Rheinland-Pfalz und Saarland. Eine geographische Landeskunde*, Darmstadt 1989.
- Hammerschmidt, H.-J., Koblenz und das Winterhochwasser 1993, *Brandschutz, Deutsche Feuerwehrzeitung*, 48, 1994, 872-877.
- Hammerschmidt, H.-J., Die beiden letzten Hochwasser in Koblenz, in: *Ratgeber. Mit dem Hochwasser leben*, 1995, pp. 56-61.

- Harreiner, J., Elektrische Felder in Netzteilnähe bei Hochwasser, *Brandschutz, Deutsche Feuerwehrzeitung*, 50, 1996, 123-126.
- Hartl, P., Einsatzerfahrungen im Einsatzabschnitt Altstadt/Innenstadt, *Brandschutz, Deutsche Feuerwehrzeitung*, 48, 1994, 858-860.
- Hartl, P., Besondere Einsätze, *Brandschutz, Deutsche Feuerwehrzeitung*, 48, 1994, 861-864.
- Heidbrink, G., Hochwasserschadenversicherung, in: *Ratgeber. Mit dem Hochwasser leben*, 1995, pp. 86-90.
- Hochwasser und seine Ursachen, Auszug aus dem Hochwasserschutzkonzept der Stadt Köln, *Brandschutz, Deutsche Feuerwehrzeitung*, 50, 1996, 106-110.
- 'Hochwasserbericht 1993/94', *Hochwasserbericht der Stadt Köln vom 24.01.1994*, Hochwasserschutzzentrale der Stadt Köln, Köln 1994.
- 'Hochwasserbericht 1995', *Hochwasserbericht der Stadt Köln vom 20.03.1995*, Hochwasserschutzzentrale der Stadt Köln, Köln 1995.
- Hochwasserschutzkonzept der Stadt Köln, *Das Hochwasserschutzkonzept der Stadt Köln*, Auszüge, Stadt Köln, Köln 1995.
- Jakob, K., *Entfesselte Gewalten: Stürme, Erdbeben und andere Naturkatastrophen*, Basel/Boston/Berlin 1995.
- Keller, R., *Hydrologischer Atlas der Bundesrepublik Deutschland*, Boppard 1979.
- Kleeberg, H.-B, O. Niekamp, *Nationalreport Hochwasser*, Neubiberg: IDNDR 1993.
- Korn, A., Staatliche Finanzhilfen, in: *Ratgeber, Mit dem Hochwasser leben*, 1995, pp. 91-101.
- Landesamt für Wasserwirtschaft/Rheinland-Pfalz, *Das Hochwasser im Januar 1995*, Hochwasserablauf und Hochwassermeldedienst, Mainz 1995.
- Licht, A., Bericht der Enquete-Kommission 'Verbesserung des Schutzes vor Hochwassergefahren', Drucksache 12/7090, Landtag Rheinland-Pfalz-12, Wahlperiode, Mainz 1995.
- Linke, B., Haftungsfragen nach Hochwasserschäden, in: *Ratgeber. Mit dem Hochwasser leben*, 1995, pp. 82-85.
- Luhmann, N., *Soziologie des Risikos*, Berlin und New York 1991.
- Luhmann, N., Risiko und Gefahr, in: *Soziologische Aufklärung, 5. Konstruktivistische Perspektiven*, 2. Aufl., Opladen 1993.
- Maurer, K., B. Fritzen, Führungsorganisation, Informations- und Kommunikationstechnik, *Brandschutz, Deutsche Feuerwehrzeitung*, 48, 1994, 847-851.
- McNeill, W., *The Global Condition. Conquerors, Catastrophes, and Community*, New Jersey: Princeton 1992.
- Menzel, L., Versorgungseinsatz, *Brandschutz, Deutsche Feuerwehrzeitung*, 48, 1994, 852-853.
- Münchner Rück, *Die Versicherung weiterer Elementargefahren in der allgemeinen Sachversicherung in Deutschland*, München 1993.
- Neuhoff, S., Weihnachtshochwasser 1993 – Hochwasser und Hochwasserschutz in Köln, *Brandschutz, Deutsche Feuerwehrzeitung*, 48, 1994, 842-846.

- Oelmann, H., Das Kölner Hocwasserkonzept, in: Stadt Köln (Hrsg.), *Abwasserforum Köln*, 4. Ausgabe Dezember 1995, Köln 1995, pp. 12-18.
- Platt, K., M. Dabag, *Generation und Gedächtnis. Erinnerungen und kollektive Identitäten*, Opladen 1995.
- Profeld, H.-J., Die Hochwasserkatastrophe 1882 in Köln, Frankfurt am Main, Mainz und anderen Städten, *Brandschutz, Deutsche Feuerwehrzeitung*, 50, 1996, 154-160.
- Quarantelli, E.L., Ausbildung – Eine rentable Maßnahme der Katastrophenbekämpfung, in: *'Impulse'-Fachinformation (DRK) für Führungskräfte*, Heft 3, Bonn 1989.
- *Ratgeber. Mit dem Hochwasser leben*, Dokumentation der Fachtagung 'Mit dem Hochwasser leben', mit begleitender Fachmesse veranstaltet von der Steinakademie in Koblenz am 10. und 11. Oktober 1995
- Rother, K.-H., Leitlinien für einen zukunftsweisenden Hochwasserschutz, in: *Ratgeber, Mit dem Hochwasser leben*, 1995, pp. 19-23.
- Schaa, W., Anleitung zur Deichverteidigung, *Brandschutz, Deutsche Feuerwehrzeitung*, 50, 1996, 127-130.
- Schaaf, O., Umsetzung des Hochwasserschutzkonzeptes Köln im Bereich der Stadtentwässerung, in: Stadt Köln (Hrsg.), *Abwasserforum Köln*, 4. Ausgabe Dezember 1995, Köln 1995, pp. 32-36.
- Schäfer, H., Weihnachten 1993. Hochwasser am Rhein, an der Mosel und Saar – Erfahrungen und Folgerungen –, *Brandschutz, Deutsche Feuerwehrzeitung*, 48, 1994, 838-841.
- Schöttler, H. (Hrsg.), *Katastrophenschutzergänzungsgesetz 1990; Ein neues Kapitel im Bevölkerungsschutz*, Regensburg 1991.
- Schöttler, H., Katastrophenschutz und Hochwasser (Teile 1-3), *Notfallvorsorge. Zeitschrift für Katastrophenvorbeugung und Gefahrenabwehr*, Heft 3/1995, Bonn.
- Seel, H.-J., D. Ranft, Hochwasserschäden bei Pumparbeiten in überfluteten Gebäuden, *Brandschutz, Deutsche Feuerwehrzeitung*, 50, 1996, 111-115.
- Seel, H.-J., Hochwassereinsätze sinnvoll ausgewertet, *Brandschutz, Deutsche Feuerwehrzeitung*, 50, 1996, 137-138.
- Spieker, H., *Novellierung der Zivilschutzgesetzgebung, – Gutachten für Hilfsorganisationen und Hilfswerke –*, Bochum 1995.
- Stallings, R.A., *Promoting Risk: Constructing the Earthquake Threat*, New York 1995.
- Steenbock, R., Die Hochwassernotgemeinschaft Mittekrhein e.V., in: *Ratgeber. Mit dem Hochwasser leben*, 1995, pp. 73-74.
- Thon, R., Dämme gegen Rheinfluten – baulicher Hochwasserschutz in Köln, in: Stadt Köln (Hrsg.), *Abwasserforum Köln*, 4. Ausgabe Dezember 1995, Köln 1995, pp. 20-31.
- Umweltbundesamt, *Hausgemachte Überschwemmungen. Maßnahmevorschläge zur Vorsorge gegen zukünftige Hochwasserschäden*, Berlin 1994.

- Vesper, J., Feuer: Lagerhaus mit Chemikalien, *Brandschutz, Deutsche Feuerwehrzeitung*, 48, 1994, 865-869.
- Vogt, R., Hochwasser in Köln. Erfahrungen, Maßnahmen, Schlußfolgerungen, in: *Ratgeber. Mit dem Hochwasser leben*, 1995, pp. 48-55.
- Vogt, R., G. Weber, J. Feyrer, Planungen für den Hochwasserschutz am Beispiel der Stadt Köln, *Brandschutz, Deutsche Feuerwehrzeitung*, 50, 1996, 116-122.
- Weißbuch, *Zur Lage und Entwicklung der Bundeswehr*, Im Auftrage der Bundesregierung herausgegeben vom Bundesminister der Verteidigung, Bonn 1995.
- Wiedemann, P.M., *Öffentlichkeits-Arbeit bei Krisen: Ein Leitfaden zur besseren Kommunikation*, Eschborn 1991.

Appendix 2

FLOODS IN WESTERN EUROPE 1993-1996

1993

JULY 1993

Southern France (Beaujolais area)

Two people died due to heavy lightning and bad weather conditions. Bridges were flushed away and the highway near St. Ettiene was blocked.

SEPTEMBER 1993

Southern France, Northern Italy

Four people died due to storm, heavy rainfall and floods in southern part of France. Two people were drowned in Northern Italy. Schools were closed and there was a power failure, leaving 30.000 people without energy.

SEPTEMBER 1993

Southern France, Northern Italy and Switzerland

Floods in Northern Italy and the Alpes in France and Switzerland. The regions of Piemonte, Lombardije and the Valley of Aosta were heavily struck. Six people died and citizens had to be evacuated by helicopter. In the French region Savogue, six people were drowned and hundreds citizens were evacuated.

OCTOBER 1993

Southern Europe

Heavy rains in the southern part of Europe led to serious flooding and landslides. At least four people perished. More than one thousand people who had been evacuated in September, had to leave their houses again.

DECEMBER 1993

North-Western Europe

Floodings in northern France, Belgium, Germany and the southern part of the Netherlands. The December Floods led to disruption of society and serious damage of properties.

1994

JANUARY 1994

France

Snowstorms in France caused a critical situation around the Dordogne and the Charante. In southern part of France, a second flood within three months threatened to happen.

Italy

More than 6.000 tourists were isolated in the northern part of Italy due to heavy rainfall and floods.

FEBRUARY 1994

France

Continuous rainfall led to new floods in France. The village Sall-en-Beaumont, located south of Grenoble, was buried by landslides. Four people were killed. The Camarque in the Rhone-Delta flooded again after a second dike breach within three months.

OCTOBER 1994

Greece

Heavy weather caused floods, turning the streets of Athens into rivers.

NOVEMBER 1994

France

At least six people died in the southern part of France due to floods. Villages from the Alpes till the Provence, the Languedoc and Corsica, were isolated from the outside world.

Italy

Heavy weather and floods hit. Particularly Italy was very badly struck. The authorities declared the state of emergency and announced a day of national mourning. Official figures indicated that there were 60 fatalities, 26 people missed and 63 people wounded. Most of the victims were registered in the north-western of Italy. The area of Piemont suffered the most. Initial estimates said that the total damage could amount to ten billion lire.

Spain

The region of Catalonia was hit by severe floodings.

1995

JANUARY/FEBRUARY 1995

North-Western Europe

Floods and threats of flooding in France, Belgium, Germany and the Netherlands. The threat of flooding in the Netherlands triggered the evacuation of 250.000 people.

JUNE 1995

Norway

Large parts of the valley were inundated by the flooding of the river Glomma. Melting snow and heavy rainfalls lead to most serious floodings in 125 years. The floods forced thousands of inhabitants to leave their homes.

1996

JANUARY 1996

France

Floodings hit the region of Herault.

FEBRUARY 1996

Spain

The southern and western parts of Spain were badly hit by heavy rainfall and snow. Especially the regions of Huelva and Sevilla were struck by large floodings. At least ten people were killed due to the bad weather.

JUNE 1996

Italy

Heavy weather in Tuscany led to eleven fatalities. Approximately 25 people were found to be missing and emergency people feared for more victims. Most of the fatalities were registered around cities of Lucca and Massa, both popular tourist centres. Hundreds of people were forced to leave their houses.

AUGUST 1996

Spain

Flash floods in Spain caught tourists in a camping site near Biescas by surprise. Eventually, 83 victims were registered and five people were found to be missing. As a result of the floods, questions were raised about the causes to be taken and the prevention of these kind of floods in general.

LIST OF CONTRIBUTORS

Michel Bezuyen is a researcher/consultant at the Crisis Research Center in The Hague, The Netherlands

Wolff R. Dombrowsky is co-director of the Disaster Research Unit at the Christian Albrechts Universität in Kiel, Germany

Menno van Duin is co-director of the Crisis Research Center, associate professor in the Department of Public Administration at Leiden University and affiliated with the Dutch Fire Service Academy, The Netherlands

Claude Gilbert is research director at the Centre National de la Recherche Scientifique (CNRS) in Grenoble, and coordinates CNRS's national research program on Collective Risks and Crisis Situations, France

Cécile Gouy is an independent researcher who regularly participates in case studies of risk prevention and crisis management in France

Paul 't Hart is associate professor and research director at the Crisis Research Center, Department of Public Administration, Leiden University, The Netherlands

Hugo van Hassel is professor of public administration and director of the Vervolmakingscentrum voor het Overheidsbeleid, Leuven University, Belgium

Leen van Lindt is researcher at the Vervolmakingscentrum voor het Overheidsbeleid, Leuven University, Belgium

Lutz Ohlendieck is an independent researcher formerly affiliated with the Disaster Research Unit at the Christian Albrechts Universität in Kiel, Germany

Uriel Rosenthal is founder and co-director of the Leiden/The Hague Crisis Research Center, professor of public administration at Leiden University and Dean of the Netherlands School of Government, The Netherlands

Marie-Louise de Vreeze-Verhoef is a research associate at the Department of Public Administration of Leiden University and affiliated with the Crisis Research Center, The Hague, The Netherlands

Addresses

Dr. C. Gilbert (Directeur de recherche au CNRS)
Cécile Gouy (Chargée d'études contractuelle (CNRS)
Centre National de la Recherche Scientifique
Programme Risque Collectifs et Situations de Crise
MRASH-UPFM-BP 47X
38040 Grenoble CEDEX9
France

Prof.dr. H. van Hassel
L. van Lindt
Vervolmakingscentrum voor Overheidsmanagement en -Beleid
Public Management Centre
E. van Evenstraat 2a
B-3000 Leuven
Belgium

Prof.dr. U. Rosenthal, dr. P. 't Hart, dr. M.J. van Duin, drs. M.J. Bezuyen, drs. M.L. de Vreeze-Verhoef
Crisis Research Center
Leiden University
PO Box 9555
2300 RB Leiden
The Netherlands

Dr. W.R. Dombrowsky
L. Ohlendieck
Katastrophenforschungstelle
Institut für Soziologie
Christian-Albrechts-Universität zu Kiel
Olshausenstraße 40
D-24098 Kiel
Germany

MIX
Papier aus verantwortungsvollen Quellen
Paper from responsible sources
FSC® C105338

If you have any concerns about our products,
you can contact us on
ProductSafety@springernature.com

In case Publisher is established outside the EU,
the EU authorized representative is:
Springer Nature Customer Service Center GmbH
Europaplatz 3, 69115 Heidelberg, Germany

Printed by Libri Plureos GmbH
in Hamburg, Germany